T0222259

OPTIMIZING DATA-TO-LEARNING-TO-ACTION

THE MODERN APPROACH TO CONTINUOUS PERFORMANCE IMPROVEMENT FOR BUSINESSES

Steven Flinn

Apress®

Optimizing Data-to-Learning-to-Action: The Modern Approach to Continuous Performance Improvement for Businesses

Steven Flinn
Brenham, Texas, USA

ISBN-13 (pbk): 978-1-4842-3530-0   ISBN-13 (electronic): 978-1-4842-3531-7
https://doi.org/10.1007/978-1-4842-3531-7

Library of Congress Control Number: 2018939149

Managing Director, Apress Media LLC: Welmoed Spahr
Acquisitions Editor: Susan McDermott
Development Editor: Laura Berendson
Coordinating Editor: Rita Fernando

Cover designed by eStudioCalamar

Cover image designed by Freepik (www.freepik.com)

Distributed to the book trade worldwide by Springer Science+Business Media New York, 233 Spring Street, 6th Floor, New York, NY 10013. Phone 1-800-SPRINGER, fax (201) 348-4505, email orders-ny@springer-sbm.com, or visit www.springeronline.com. Apress Media, LLC is a California LLC and the sole member (owner) is Springer Science + Business Media Finance Inc (SSBM Finance Inc). SSBM Finance Inc is a **Delaware** corporation.

For information on translations, please email rights@apress.com, or visit http://www.apress.com/rights-permissions.

Apress titles may be purchased in bulk for academic, corporate, or promotional use. eBook versions and licenses are also available for most titles. For more information, reference our Print and eBook Bulk Sales web page at http://www.apress.com/bulk-sales.

Any source code or other supplementary material referenced by the author in this book is available to readers on GitHub via the book's product page, located at www.apress.com/9781484235300. For more detailed information, please visit http://www.apress.com/source-code.

Printed on acid-free paper

To my mother and father

Contents

About the Author

Steven Flinn is founder and CEO of ManyWorlds, Inc., which is a pioneer of machine learning–based solutions for enterprises, the market-leading provider of visual UX software for collaborative systems, and a provider of related advisory services to leading organizations around the world. Mr. Flinn has extensive consulting experience at the intersection of strategy, decision science, and technology with Global 1000 enterprises, as well as with selected high-impact startups. He has been awarded over 40 patents in the field of machine learning and its applications and is the author of *The Learning Layer* (Palgrave Macmillan, 2010), which predicted, and established the imperative for, applying machine learning–based capabilities in the enterprise, an imperative that is now widely accepted and a reality. Prior to ManyWorlds, he was a chief information officer and vice president of strategy at Royal Dutch Shell. His education includes graduate degrees from Northwestern University's Kellogg School of Business and Stanford University's School of Engineering.

About the Technical Reviewer

Sébastien Caron is the founder of Conova Solutions, a management consultancy and strategic advisory firm that helps organizations leveraging the digital workplace to attract talent and build in their DNA the capabilities required to address increasing complexity and innovate continuously. His expertise covers approaches and models such as co-innovation, internal lean startup, social business, Agile, Cynefin model, SECI model, capability model, maturity model, and design thinking. He has worked with organizations such as CGI, Loto-Québec, and recently with the National Bank of Canada to help them design and implement an insight-driven organization (enterprise knowledge graph, machine learning, analytics, personalized recommendations, and virtual assistant). As a PhD researcher in cognitive and computer science, his expertise has been recognized by many awards: winner of the Cognitive Informatics Symposium contest in Montreal (2011 & 2004), winner of the International Scientific Competition of Ile de France Regional Council (2005), winner of the International Competition for Researchers in Information Science & Computing Technology (2005), and more. He likes doing robot fights with his kids, playing soccer during the summer, and killing zombies in virtual reality games with his friends.

Acknowledgments

This book synthesizes concepts that I have in recent years developed and put into practice that are, in turn, based upon experiences and learnings spanning a good number of years before. For example, I am grateful for obtaining an excellent grounding in mathematics and economics as an undergraduate at Binghamton University. I was fortunate to have benefitted from Dr. Ronald Howard's pioneering work in the field of decision analysis during my graduate studies in Stanford University's Engineering-Economic Systems Department. I was also fortunate to attend Northwestern University's Kellogg School of Management and benefit from, among other things, a deep dive into strategy and finance. And at Royal Dutch Shell I had the opportunity to experience the excitement and challenges of leading information technology at the global-enterprise level. Finally, at my current company, ManyWorlds, I have been able to pioneer machine-learning applications for enterprises, as well as to advise, and learn from, a remarkable mix of leading businesses on a variety of subjects at the intersection of strategy, decision science, and information technology. It seems as if this unique melange of opportunities and experiences was necessary to make the method outlined in this book fully come together. I very much doubt it could have happened otherwise.

In making the book a published reality, I'd like to thank the wonderful editors at Apress, particularly Susan McDermott and Rita Fernando, for believing in the work and keeping everything on track. Thanks to Naomi Moneypenny for pioneering with me at ManyWorlds many of the concepts that are detailed in the book and providing early guidance on the manuscript. And a very special thanks to Sébastien Caron, who agreed to be the technical reviewer for the book (without knowing exactly what he was getting into!), helping me to see some of my blind spots and making the book better than it would otherwise be. Of course, any deficiencies of the book remain solely my responsibility. And finally, thanks to my family for being understanding about the commitment (which is always greater than expected!) that writing a book requires.

Introduction

We live in interesting times. On the one hand, technology-based advances are occurring at a bewildering pace. But on the other hand, a careful examination of macro-level data suggests that U.S. business performance continues to flag compared to historical results. And it seems that our current approaches to improving business performance remain stuck in the last century. If the technology references are stripped out, one would be hard-pressed to distinguish the popularly prescribed methods in today's management publications from those of a decade or two ago. Really, not much seems to be new in that regard since at least the 1990s.

This book aims to challenge this business-performance somnolence by putting forward a new approach to performance improvement that is relevant for *today's* dynamic environment and promises to be robust enough to continue to stay relevant. The design goals of this new approach are as follows:

- Provide a clear and effective method for continuous performance improvement for today's organizations

- Demonstrate how to systematically get the most out of people, process, and technology investments, and do so on a continuing basis

- Be widely applicable to different business sectors and business functions

So, What Is It?

This new approach begins with a recognition that the key to improved business performance in today's world is *improved decisions*. That really has always been the case, but this reality has often been skirted past or treated in a tangential way by the continuous stream of popular management memes. And the data certainly backs up the intuition that if you improve decisions you improve financial performance. Bain & Company, for example, found that "decision effectiveness and financial results correlated at a 95% confidence level or higher for every country, industry, and company size."[1]

---

[1]Blenko, Marcia, Michael Mankins, and Paul Rogers, "The Decision-Driven Organization" *Harvard Business Review,* June 2010. https://hbr.org/2010/06/the-decision-driven-organization

And not all decisions are created equal—it is the decisions that are aligned with the most important drivers of value for an organization that really make a difference, and those drivers may not always be obvious. It always has been the case, although often not explicitly recognized, that the bigger impact with respect to the long-term value of, for example, a manufacturing facility, is not necessarily the way it is operated at any given time. Rather, it is more likely the decisions on whether it should be built at all, how it should be configured, where it should be located, what logistics should be developed to support it, and so on, that will dictate the enduring performance legacy.

The field of decision analysis, a prescriptive discipline for improved decision making that has evolved over the past half century, can classically be applied to help with big, one-off, decisions such as making an acquisition or whether and where to build a plant. But it has been somewhat awkward to apply full-blown decision analyses to those myriad recurring decisions that are at the heart of what organizations do on a day-to-day basis. The approach presented here uniquely adapts key aspects of decision analysis and integrates them with techniques from other fields of management science so that they can be beneficially applied to *all* types of decisions that are made in organizations, and at all levels of the organization.

By putting decisions front and center, we can more readily dispense with the fuzziness in thinking that often comes along for the ride with the viral popularity of the latest technology and management memes. For example, take that field of exploding popularity, data science. The question to consider for any set of "big data" or associated data analysis is, "What decisions can potentially be improved, and what is the financial upside for that improvement?" Those new machine learning–based capabilities that are being breathlessly touted? Same question. A new system to better facilitate collaboration? Same question. A new marketing process? Same question. You get the picture. It seems so simple, and yet these questions are often not asked, and if they are, they are not answered in any rigorous way.

The second foundational element this new approach emphasizes is *learning,* in its most robust meaning and application. It is subordinate to decisions in that learning—at least in the business sense, as perhaps opposed to our everyday life—*only has value insofar that it has the potential to change a decision that would otherwise occur,* a decision that in turn would change an action that would otherwise occur. And the corollary is that data, and its more filtered and organized derivatives, information and knowledge, only have value insofar as they enhance learning. This is an important concept: there is no *intrinsic* value to data (or information or knowledge) unless it serves as a basis for enhanced learning or has the potential to do so. Classic decision analysis has a concept of "value of information"—and we will discuss and apply the concept in this book—but the label is somewhat of a misnomer. What that notion really connotes is *value of learning,* and data, information, and knowledge only have value to the extent that they are derived from this value of learning.

Of course, speaking of learning, in today's world there has emerged an entirely new aspect of learning that never needed to be seriously considered even as recently as just a couple of years ago: *machines that learn*. In the past, we thought of learning as only a human endeavor, and the focus of the field of organizational learning was obviously and understandably focused on human learning. But going forward, we need to increasingly consider both human and machine-based learning in just about any organizational process or activity, and our more robust approach to learning in this book seamlessly extends to machines as well as people.

What Is It Not?

Optimizing data-to-learning-to-action processes does not somehow make obsolete the other valuable disciplines of business excellence and improvement; rather, it generally complements them and amplifies their value. It does not replace business strategy, for example, but it can inform strategy and help you get the most out of your strategy. It does not replace classic initiatives such as, for example, quality improvement, but it can provide guidance on where and when to invest in such initiatives. It does not get in the way of innovation, but it can help make innovation-based processes more efficient as well as effective. While there is necessarily a significant emphasis on technology applications and quantifications in the optimizing of data-to-learning-to-action, it is the classic trifecta of people, process, and technology, in that order of importance, that ultimately rules. What the method presented here can do is help provide an insightful rigor on where, when, and how to emphasize these universal business assets and disciplines in your organization.

Why Now?

First, because business performance needs to improve and the classic answers of the past few decades have run out of steam for many organizations. Quality improvement? Sure, but that has already been widely applied (and if not, should be!). Reengineering? Great, but by now it has basically run its course in most organizations and typically has focused on limited parts of the organization that are more transactional in nature. Leadership, coaching, organizational design, change management, etc.? Absolutely, these disciplines will always be an important focus, but they are not sufficient to achieve and maintain the most outstanding performance. That requires new and superior methods and practices, and that's what this book aims to deliver.

Second, the economy is rapidly transforming toward most work processes becoming knowledge, learning, and decision-intense, while areas of business that are not knowledge, learning, and decision-intense are quickly becoming automated and routine. That means the methods and practices that really matter must increasingly focus on the knowledge, learning, and decision-intense areas of the organization, which is exactly what the approach advocated by this book does.

Third, because the pace of advances in technology, particularly information technology, is accelerating—including information technology that is applied in the enterprise—businesses need a way to make the most of these advances, and to do so efficiently. Artificial intelligence broadly, and machine learning specifically, may well be the most important of these accelerating technologies, but concurrently we have continuing advances in areas such as the internet of things, new ways for people to collaborate, business intelligence, and the movement of enterprise IT in general to the cloud. The latter is an accelerator for everything else because of the cloud's ability to deliver advances in capabilities and features almost immediately. There is simply no overstating how much of an impact these technologies will have on businesses over the next decade and beyond. But we also know very well from previous technology inflection points that there will be plenty of potential money pits and blind alleys along the way, so there is also no overstating how difficult it is to make optimal decisions about when to invest in these technologies, how to justify the investment, and how and where to deploy the technologies in an organization. And that's why answering these technology-related questions is a key focus of this book.

Why Should It Matter to Me?

If you are a business leader, this approach should matter to you because it is up to you to ensure that your organization is doing everything possible to optimize its performance and to continuously maintain that outstanding performance. It is also up to you to ensure that investment decisions are backed by rigorous justification—you have a fiduciary responsibility to do so!

If you are an IT executive, it should matter to you because you need a method that can add additional structure to the technology-related investment and deployment decisions you make on a continuous basis, and you need to have a solid way to justify your recommendations to your leadership and business colleagues.

If you are technology practitioner or a knowledge worker, you should care because it will make life much easier for you when making the countless daily assumptions and decisions that are a natural part of your work by providing a guiding framework for them that you can confidently rely on.

If you are a technology provider or provide related services, it should matter to you because the methods presented here can provide a means for you to demonstrate where and how your product and services can deliver value within your clients' data-to-learning-to-action processes and support credible quantified estimates of the value they deliver.

If you are a management consultant, you should care because the approach and methods described in this book are widely applicable and therefore offer an opportunity for you to build a very significant new practice!

And if you are a data scientist and/or machine-learning expert (lucky you!), you should care because the approach in this book will enable you to put your efforts in the context of overall business decision making and learning, making the strategic importance of your work even more clear.

Overview of the Approach

At its core, this approach is guided by and integrates four main themes:

- Optimizing decisions, particularly high-leverage recurring decisions

- Treating the data-to-learning-to-action flow in which decisions are embedded explicitly as a process

- Putting the concept of learning on a robust and rigorous foundation

- Identifying and resolving constraints on value within the data-to-learning-to-action process

Decisions are at the core of business-performance improvement—improve decisions and you improve performance. And not just any decisions, but the ones that are most aligned with an organization's important *value drivers*. As I mentioned earlier, that has always been the case, but frankly, as technology advances and more mundane tasks become automated, making decisions in complex situations is really the primary activity that is left for people to do in most parts of business! And, again while a mature field developed to address large, one-off decisions, it has been cumbersome to try to apply it to smaller and/or recurring decisions (and in practice it seems to only rarely be fully applied to the large, one-off decisions, unfortunately). In the pages of this book, we'll demonstrate how to adapt key elements of the decision sciences so that they can be effectively applied to all those other types of decisions that constitute the bulk of what we all do in the business world. The result of doing so is a rigor and clarity of thinking that has often been lacking in much of business decision making.

Decisions are also at the core of the informational flow that constitutes data-to-learning-to-action, since learning, and its inputs of data, information, and knowledge have value only by virtue of their affecting decisions. (And we'll address the relationship among data, information, and knowledge in detail in subsequent chapters.) But beyond simply representing a natural flow of information, we can also treat data-to-learning-to-action as an explicit process. In fact, it is a *universal process*. Decisions come right after the learning step and right before the action step in this data-to-learning-to-action process. That process sequence holds true for business applications, is true for our personal lives, and is even true with respect to other organisms. It is also true of machine-based learning. With apologies to Cole Porter, birds do it, we do it, even educated machines do it! It's truly a universal process—the one process that rules them all. That's because it is *the* fundamental learning process, and learning is core to our business lives as well as to intelligent actors in general. We/it learns, which informs a decision, which then drives an action. And the learning is based on experience (i.e., data). Data-to-learning-to-action is often a *recursively adaptive* process because the actions resulting from the process can create new data, which is an input to learning, which in turn informs decisions, which drive new actions, and so on.

So, data-to-learning-to-action is a process—a universal process—and, of course, processes have a flow. Our process starts with inputs or data, which may be transformed in various ways (to information and/or knowledge) and are then consumed by learning processes (yes, processes can naturally have sub-processes!). The learning is then applied by the decision process, which results in an action (whereby *action* is defined sufficiently robustly so that deferring taking an action can also be considered an action).

All well and good, but what exactly do we mean by *learning*? Ah, that is a central issue to be addressed here because the reality is that in legacy management approaches exactly what is meant by *learning* is invariably glossed over, either by tacitly assuming that learning is an intuitively obvious concept and needs no further explanation, or, perhaps more likely, resulting from being resigned to the assumption that learning is simply too slippery a concept to be amenable to more rigorous definition. That won't do for our purposes. We need to define the phenomenon known as learning precisely enough such that it can be *quantified*, else we have no way to measure and optimize the business value of the phenomenon. At the same time, the definition needs to be *fully applicable to both minds and machines*, else we do not have an approach that is sufficiently extensible in the face of the inevitable advances in machine learning and its ubiquitous application in the coming years. Fortunately, we will see that the essence of learning is the *reducing of uncertainty*, thereby enabling more accurate predictions, and that this insight on the essential nature of learning does the double duty that we need: it both enables a quantification of the value of learning and is fully applicable to both the human and the artificial versions of learning.

Because processes have a natural flow, just as with any other type of flow, there can exist constraints or pinch points that limit the flow of the output of the process, with the flow of our data-to-learning-to-action process comprising actionable, and therefore valuable, learning. In a river, the constraint may be a narrow canyon. For electrical flows, it may be a patch of high resistance. For a data-to-learning-to-action process, it may be the result of any number of constraints or bottlenecks of actionable learning along the process chain, but whatever they are, we will see that they invariably are tied to one or more people, process, or technology factors. What's different with our data-to-learning-to-action flow is that, unlike a river, it can often be tricky to quickly spot the primary bottleneck. Our process tends to be much more complex, and first impressions can easily be wrong!

But we can apply our own "trick" or alternative perspective that enables us to cut through the complexity and bring to light the constraints of our process. (And the fundamental law of flows is that there is always a constraint somewhere!) In a sense, we reverse the flow to find the constraints. We work backward from the decisions to the learning, and then back to the data, and through all the intermediate steps in between. We do that in a precise way, aligned with decision analytic–based methods, which enables us to identify one or more locations in the chain that result in a constraint to learning, and, more specifically, to the learning that really matters with respect to the decisions with which we are most concerned. In many cases, we can quantify quite rigorously the value of alleviating the constraint. This perspective of working backward along the data-to-learning-to-action flow is a core concept of the book. The book could easily have been alternatively titled *Decision-to-Learning-to-Data*!

It should be apparent that the identification of constraints in the data-to-learning-to-action flows is necessarily a continuous process. Since a fundamental property of flows is that there is inevitably a bottleneck somewhere, there will likely be a new one to be considered after the initial one is successfully addressed. But by taking this step-wise approach we avoid the wasteful over-building and bad timing that organizations are so prone to, particularly with respect to information technology.

The Flow of the Book

The flow of the book mirrors that of this introduction. We will first build out the case for action and then explore the disciplines that serve as the roots of this new approach, such as decision science, constraint theory, and the process perspective. We'll then examine the data-to-learning-to-action chain in detail and take a tour of each of the elements. During the tour, we'll consider examples of each of the elements, as well as external factors that are or will be impacting each of the elements, particularly technology-related factors. We'll also see why the "chain" is often really a closed loop. Next, we'll discuss how to work backward along the chain to identify value bottlenecks. We'll introduce the concept of the *value of learning* and provide an overview of how to rigorously and credibly quantify this value, which enables us to quantify the value of solutions to the bottlenecks. Quantifying the value of learning also enables a quantification of *total value*, which is *the* appropriate measure for prioritizing an organization's investments. We'll then spend a good deal of time on common types of bottlenecks that are associated with each individual data-to-learning-to-action element and people-process-technology examples of how the bottlenecks can be addressed. Finally, we will spend some time on how to organize for, and perform, data-to-learning-to-action optimizations that will add value for your organization.

With that, I send you on your way to a reading, learning, and, hopefully, action adventure!

Case for Action

Or more precisely, this chapter is all about the case for data-to-learning-to-action! We touched on "Why Now" in the introduction—here, we will take a much deeper dive into that subject. A fundamental maxim of change management is that organizations will not take a new direction or take on a new approach unless there is genuine dissatisfaction within the organization with the status quo situation. This chapter provides plenty of reasons why there should be cause for concern about the current state and should help serve to get the optimizing data-to-learning-to-action approach off the ground in your organization.

The following are motivators for moving beyond the current state of "business as usual" that we will explore in detail this chapter:

- A careful examination of the long-term trends of the economic results for firms, particularly in the United States, over the past several decades reveals a not-so-pretty picture, and the picture appears to generally be worsening.

- The dizzying advances in information technology, while ultimately promising a tremendous upside, also create an intensifying level of complexity and confusion, too often leading to either organizational paralysis or wasteful spending.

© Steven Flinn 2018
S. Flinn, *Optimizing Data-to-Learning-to-Action*,
https://doi.org/10.1007/978-1-4842-3531-7_1

- Along with the baseline complexity and confusion that the rapid advances in technology are leaving in their wake, the level of confusion is amplified by the advocacy of players who promote their own agendas with respect to particular technologies or technology-based roles, further serving to inhibit clear thinking about business value.

- The historical toolkit of management techniques aimed at improving business performance remains valuable, but fails to effectively address key performance-improvement issues that are relevant for organizations operating in today's environment.

The Economic Imperative

Deloitte's Center for the Edge has conducted a remarkable study over the past seven years or so that brings into sharp focus the economic issues that businesses are facing in the contemporary economic and technological environment. This study includes a periodically published set of metrics and accompanying commentary that Deloitte calls "The Shift Index."[1]

The Shift Index illustrates that, on the one hand, some US economic-performance indicators, such as productivity growth and overall GDP growth, have been increasing, and these positive indicators seemingly provide comfort with respect to overall economic performance. On the other hand, a deeper look reveals troubling systemic issues related to business performance—issues that the optimizing data-to-learning-to-action approach is geared to help address.

Perhaps most sobering is the long-term trend of the return on assets (ROA) for US firms since the mid-1960s, which is depicted by Figure 1-1, along with the associated linear trend line. The steady deterioration of ROA over the multiple decades is both obvious by inspection and alarming. Return on assets is defined in accounting terms as net income divided by assets. Basically, it can be considered a measurement of how effectively a company's assets are being leveraged for economic benefit. The assets may be hard assets, such as plants and equipment, but also less tangible items, such as software and even cash holdings. In many ways, this decline in return on assets is particularly surprising because with the increasing proportion of services in the economic mix at the expense of hard assets, it would be expected that the ROA metric would benefit. Hard assets should be proportionally shrinking, and their value-add should therefore be increasing as a result of the ever-growing levels of services

---

[1]https://dupress.deloitte.com/dup-us-en/topics/strategy/shift-index.html; https://www.forbes.com/sites/stevedenning/2016/12/15/shift-index-2016-shows-continuing-decline-in-performance-of-us-firms/#f2233b3386cc

that are not part of the asset-based denominator but that do contribute to the net income in the numerator.

But, in fact, ROA has continued to decline *in spite of the help* it should be getting from this services-to-asset mix advantage. ROA is a fundamental—perhaps *the* fundamental—way to judge the overall economic performance of businesses, and therefore this deterioration needs to be taken very seriously, notwithstanding the gloss of seemingly benign economic news embodied by other, less fundamental, metrics.[2]

So, what is the root cause of this deterioration of ROA? Most fundamentally, *the only way aggregate ROA can continue to decline is because decisions with respect to investments in, and operations of, assets are relatively poorer than they were historically.* It's as simple (and complex!) as that, since we know financial performance is highly correlated with decision effectiveness.[3] In fact, particularly puzzling is that this decline in the economic performance of businesses is occurring in the face of a concurrent explosive growth and popularity of business schools, management-related publications, and the overall field of management consulting! How can that possibly be? The inescapable answer is that business decision making, in aggregate, must somehow be worse than it was historically, *despite* the concurrent growth in management-related education and advice.

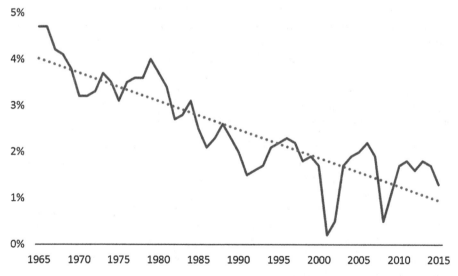

Figure 1-1. US firms' return on assets 1965–2015. Based on Deloitte/Compustat data.

[2]https://www.forbes.com/sites/stevedenning/2011/10/19/the-big-shift-or-shifty-statistics/#5c394f85674e
[3]Blenko, Marcia, Michael Mankins, and Paul Rogers, "The Decision-Driven Organization", *Harvard Business Review,* June 2010. https://hbr.org/2010/06/the-decision-driven-organization

Perhaps some insight into the paradox lies within a seemingly completely different puzzle in the field of medical diagnostics. Over the past few years there has been significant controversy about cancer screenings, particularly for prostate and breast cancers.[4] On the one hand, it has traditionally seemed sensible to encourage such screenings, even though the screens are not perfectly reliable. That is, they are prone to some degree of false positives and false negatives. Nevertheless, the screens seem to at least provide useful clues that can then be followed up on, and a fundamental maxim of decision science is that information, even if it is not perfect, cannot be worse than not having the information at all. Or can it? When the all-type mortality of those screened was compared to that of those who were not screened, it was found that those people who had been screened on average had worse outcomes than those who had not been screened![5] How could that be? How could *more* information possibly be worse than *less* information? The only way that could be the case is if the information was somehow systematically misused. That is, while the screening information should *in theory* enable better decision making, it was, in fact, causing worse decision making than if no screening tests had been conducted. Specifically, over-treatment was apparently occurring, and treatments always carry their own risks, even if the risks are comparatively small.

In other words, although with sophisticated decision making the screening information would enable better treatment decisions to be made, resulting in comparatively better all-mortality outcomes, applying unsophisticated or downright poor decision making transformed what should have been something with a positive value into something with a negative value. Unfortunately, rather than tackle the root cause of the problem, which is clearly the decision-making process occurring downstream of the screening, various medical organizations essentially threw up their hands and recommended that in most circumstances the screenings simply not be performed at all!

The lesson from this unfortunate situation is that even positive advances, if they come along with complexity and uncertainties, can lead to *value destruction* rather than value creation if the associated decision-making processes are not up to the task. From the perspective of the overall economy, decisions are currently being made in an environment that is characterized by greater complexity, more rapid change, and more new uncertainties than ever before. And that makes for very challenging decision making, necessitating more thoughtful approaches if the advances are to lead to business value creation rather than value destruction.

---

[4] See, for example, http://www.health.harvard.edu/mens-health/the-new-psa-report-understand-the-controversy
[5] http://ascopubs.org/doi/full/10.1200/jco.2015.61.6532

Decomposing the aggregate returns on assets of Figure 1-1, it is noteworthy that even for companies in the top quartile, ROA is at least modestly falling, which implies that even these top-performing firms' decisioning processes need work. And as illustrated in Figure 1-2, for the bottom quartile of companies, sub-par decision making results in ROA levels that are prone to being negative and that plunge dramatically during times of overall economic stress.

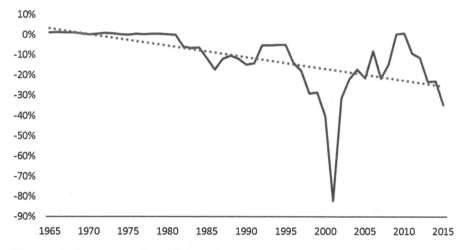

Figure 1-2. Bottom quartile of US firms' return on assets 1965–2015. Based on Deloitte/ Compustat data.

Chronically low returns on assets result in more than just ugly-looking accounting charts. In times of stress, it is the path to dramatic changes in competitive fortunes, perhaps even leading to business extinction events. This is well illustrated by Figure 1-3, which charts the "Topple Rate," a metric of changing ranks among companies with greater than $100 million in annual revenue.[6] Bucking the long-term trend, there has recently been a pause in the increasing rate of rank churn. The long-term trend of the topple rate suggests that we may merely be in a period of calm before the storm in that regard, however. For example, as we saw in Figure 1-2, ROA has continued to decline during this same recent period. It seems a reasonable conjecture that we are therefore set up for another spike in toppling as soon as the relatively benign economic period of the last few years inevitably comes to an end.

---

[6]For more information on topple rate, see https://hbr.org/2005/03/the-faster-they-fall

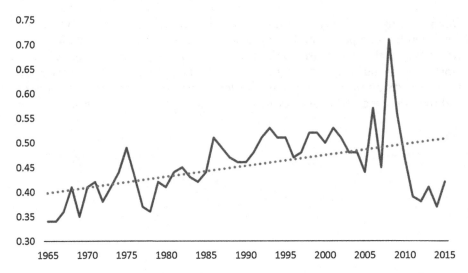

Figure 1-3. US company topple rate 1965–2015. Based on Deloitte/Compustat data.[7]

Other metrics associated with the Shift Index are broadly consistent with the unfortunate trends of return on assets and topple rates. Take, for example, the financial metric that ultimately matters most to the owners of enterprises—return on shareholder value. While rates have been maintained by the top quartile of performers, for the bottom quartile, shareholder value is being destroyed, and seemingly at an increasingly faster rate.

Ongoing inadequate returns on assets and the destruction of shareholder value can only occur when enterprise decision processes are suboptimal or, in the worst cases, just plain consistently faulty since, "ultimately, a company's value is just the sum of the decisions it makes and executes."[8] And, of course, decisions tend to be suboptimal or faulty because learning is suboptimal or faulty. Which again leads us to the inevitable conclusion that optimizing data-to-learning-to-action is *the* way, and really the *only sustainable* way, to get out of the rut of sub-standard business performance.

[7]Deloitte's analysis was further based on Thomas C. Powell and Ingo Reinhardt, "Rank friction: An ordinal approach to persistent profitability", *Strategic Management* 31(11), November 2010, 1244–55.
[8]Blenko, Marcia, Michael Mankins, and Paul Rogers, "The Decision-Driven Organization", *Harvard Business Review*, June 2010. https://hbr.org/2010/06/the-decision-driven-organization

Disruptive Technologies

As is the case in nature, disruptive forces and the resulting stress are the drivers of extinction events in the business world. Economic recessions are classic stressors that serve to clear out the corporate old, the young, and the infirmed, and that reality is reflected in some of the abrupt spikes and deep valleys depicted in the business-performance charts we just reviewed. But sometimes even forces that would seem on balance to be good things can be stressors as well, because they can mean significant change is required. And change, even when the change seems destined to result in a positive outcome, is most definitely an organizational stressor.

The rapid advances in technology we are currently experiencing represent just such outwardly attractive opportunities, but also represent potentially fatal stressors for organizations. The list of disruptive technology and related advances that enterprises now need to successfully navigate is formidable and just keeps growing. It can seem like a big, buzzing confusion, as illustrated by Figure 1-4.

Figure 1-4. IT-driven opportunity and confusion

When there is a buzzing confusion, inevitably decisions are going to be suboptimal. Fear, uncertainty, and herd mentality all play their part in sabotaging logical thinking. When whimsical decisions about information technology, data, and analytics are continuing favorite topics of Dilbert cartoons, you know there is a real problem!

The following is a brief look at just a few of the potentially disruptive technologies and related trends that are profoundly affecting today's organizations, for better or worse.

Cloud Computing

The inevitable transition to cloud computing models brings with it significant efficiency benefits. Fixed costs are converted to variable costs. Upgrading to new features is a much more graceful process. There is less friction in accessing and sharing information. New capabilities can be delivered to users much more quickly.

Cloud computing has already disrupted entire sectors of the economy. Information technology companies themselves have been particularly affected. The model has enabled new entrants to quickly gain traction in the marketplace at the expense of traditional on-premises vendors. And incumbent software companies have had to meet this competitive threat by undertaking the onerous task of transforming their product lines from on-premises to cloud-first models, while navigating a revenue-model shift from up-front licensing fees to a subscription model. So, cloud computing–based decisions are increasingly fundamental to competitive-positioning decisions in some sectors of the economy—and getting it wrong can be fatal.

But cloud computing is also at the core of most decisions with respect to the internal information technology of businesses in general. These decisions are a fundamental part of any CIO's job these days. Moving to the cloud brings with it all those tremendous advantages for the organization just described: upgrading to new features is much easier, and new capabilities can be delivered to users much more quickly—and can be more easily accessed by users without the IT organization's involvement (which, of course, brings with it both positive and negative aspects).

Although these are significant advantages, they can be organizational stressors as well. It must be carefully thought through how new cloud-delivered capabilities are applied in practice. If they are applied haphazardly or without thorough thought of how they will integrate with the current technological or process environments and how they will integrate across cloud platforms, as well as how the associated changes will affect user communities, they can surely lead to negative rather than positive value. Cloud computing is the *amplifier* of the advantages that other technologies promise, but also the amplifier of the buzzing confusion that accompanies all the new technology options.

Internet of Things

The internet of things (IoT) denotes an intensely networked world in which every device is connected: clocks, refrigerators, vehicles, industrial sensors, medical implants, and so forth, in addition to the already ubiquitous smartphones. And it is a world awash in all the data that emanates from all these devices.

The IoT phenomenon affects every company that delivers "things" into the marketplace. The things must be increasingly intelligent and connected. In some cases, they may be self-propelled and embodied as a robotic apparatus. Difficult decisions must be made between the legacy things and these new IoT-based products. As in the case of IT vendors with respect to cloud computing, getting this transition right is a life-or-death decision process for product companies.

And for businesses that are the consumers of the products, determining when to convert from legacy equipment to smart, connected devices and equipment is difficult. There are not only timing considerations, but also integration considerations and, of course, an overlay of security and privacy issues to be considered as well. Simply being connected to the internet is only half the story. For example, how are the new devices integrated into an organization's overall processes and IT infrastructure? Cross-vendor compatibility is always an issue. Is there a special-purpose application programming interface (API) for the new device, or does it obey an industry standard API? And how long can the standard be expected to hold?

In summary, the internet of things promises wonderful new opportunities for businesses. But it also contributes to the buzzing confusion that makes decisions that much more difficult by creating an uncertainty about just when the right time is to convert to the new devices, by creating potential standards and compatibility complexity and confusion, and by strongly contributing to the next huge opportunity, but also decision menace, on our list—"Big Data."

Big Data and Business Intelligence

Has anything been hyped over the past few years as much as big data has? Or its close cousin, data science? Of course, big data is not really a technology per se, but the concept fits with technology-related stressors, so I'll touch on it here, as well as the general term that is often applied to the technologies that facilitate analysis of all that data, *business intelligence*.

In this book, we will continuously hammer home the point that it is *decisions* that matter—data only has value to the extent that it has at least the potential to affect a decision. And data can only affect a decision by affecting learning (and that learning and the resulting affected decisions may now very well be made by a machine rather than a human, as will be discussed in the next section).

Nevertheless, it is very much the case that more data is being generated now than ever before, the rate of generation is ever increasing, and this super abundance of data can provide amazing benefits. Marketing processes, for example, are being transformed by the massive amount of data that is generated through the capturing of all manner of consumer behaviors. Connected devices generate real-time data that is revolutionizing manufacturing, supply chain, and medical processes, for example. Data itself has truly become a disrupter!

But in each of these cases, the fundamental question that must be asked is, "What decisions can be improved by the data?" Answering that question enables the prioritization of exactly what data to acquire, how good the data needs to be, what the learning objective should be with respect to the data, what data science, knowledge management (KM), and/or business intelligence tools and techniques should be applied to support the organizational learning objectives, and so forth. Without that perspective, it is very easy to wander around lost in the trees because you don't have an adequate perspective on the overall forest. That leads to a loss of perspective on what really matters, or analysis paralysis, and that leads to value destruction rather than value creation.

Machine Learning/AI

Surely the most important technology trend of our time is the rise of machine learning, along with the related broader concept, artificial intelligence. I wrote a book on the promise and the imperative of machine learning in the enterprise some eight years ago, and the future I described has now definitely arrived.[9] But the surface has only been scratched. For one thing, even just those few years ago, the revolution in machine learning that has been engendered by more effectively applying neural networks, particularly deep learning–based neural networks, had not yet even really begun. And now we not only have deep-learning neural networks, but also neural network–based systems that can teach themselves, at least in specific domains, to rapidly achieve capabilities beyond those of any human![10]

Essentially, machine learning is an automation of our universal process of data-to-learning-to-action. And because of this universality, machine-based learning promises to become a ubiquitous capability. It will be embedded in nearly every device. The internet of things will surely actually be the internet of auto-learning things.

---

[9]*The Learning Layer: Building the Next Level of Intellect in Your Organization* (Palgrave MacMillan, 2010)

[10]https://www.technologyreview.com/s/609141/alphago-zero-shows-machines-can-become-superhuman-without-any-help/

But machine learning, and AI in general, is an example of technologies in which a danger for decision making is in succumbing to over-optimism in the near term, but having insufficient optimism and consideration of the art of the possible in the longer term.[11] Hype creates an environment in which bad decisions are made. On the other hand, complacency in the face of inevitable trends leads to lagging competitiveness or worse. So, timing is everything for the difficult decisions with respect to machine-learning capabilities—it must be gotten right.

And it is not just decisions on *whether* to apply machine-learning capabilities or not. Importantly, it is in *what ways* to apply machine learning. Machine learning has two overarching objectives in business applications: 1) to make people more productive by learning from experiences with them and thereby more intelligently personalizing interactions, including making people more aware of opportunities for self-improvement so that they are more productive, and 2) to do better or faster or more inexpensively by machine activities that people currently perform. Bluntly, it comes down to whether to position machine-learning capabilities to assist people or to replace people. Over time, in many application areas, the latter applications will inevitably consume the former applications, which makes investment and deployment decisions even trickier.

Enterprise Collaboration

How people work in organizations is transforming. Only a few short years ago, email and face-to-face meetings were the predominant means of interacting. The adaptation of consumer social networking and chat to the enterprise is rapidly changing that, enabling people to much more effectively collaborate on projects (e.g., applying Agile methods) and specific work products, to leverage the expertise of colleagues who they may not already know (e.g., through communities of practice), to more quickly make collective decisions, and to promote a greater sense of belonging and connection, as well as cooperation, among members of the organization.

On the other hand, this transition from an email-centric world causes stress and potentially even loss of productivity for those who are less able or willing to adapt to these new ways of working. And given the many different options and variations available, the advances in the enterprise collaboration space also generate inevitable questions on exactly what applications should be used and when for various communications and collaboration purposes. That confusion can lead to decision thrashing and fragmentation that serve to sabotage the benefits that these new technologies would otherwise deliver.

---

[11]See Amara's Law: Susan Ratcliffe, ed. (2016). "Roy Amara 1925–2007, American futurologist", *Oxford Essential Quotations* (4th ed.), Oxford University Press.

The Combinatorial Effect

We have touched on just a few of the beneficial technologies that can deliver enormous benefits, but may also sow confusion and end up being counterproductive if not acquired, deployed, and managed with care. Most of these technologies were not even on the agendas of CIOs and other executives just a few short years ago. Given that, we should expect that there will be even more exciting technologies establishing themselves in the next few years that will add to the complexity and disruption. An obvious example is blockchain-based technologies and processes. Virtual reality and "mixed reality" technologies, as well as 3D printing, are other examples. And even in some of the areas that we have already discussed, such as AI/machine learning, subfields are rapidly emerging that require independent consideration, such as the rise just in the past year or so of intelligent chatbots that are beginning to transform the default way we interact with computing systems.

It is very often the *combination* of existing and emerging technologies that drives the greatest benefits, as well as transformations, in organizations. The multiple rapidly advancing and evolving technologies of our era will surely combine into integrative opportunities that are difficult to predict but that will undoubtedly have significant business impact. For example, the opportunities that are afforded by combining business intelligence and advanced machine learning. Or the integration of mixed-reality technologies and ecommerce. The combinatorial effect of technologies, particularly those driven by that universal combining agent, machine learning, will ensure that virtually no part the enterprise will be immune from transformational pressures over the next decade.

The Confusion

The word *unprecedented* is often overused, but its use with respect to the current period of IT advances, churn, and upheaval I submit is clearly warranted. As we have briefly reviewed already, and with just the sampling illustrated by Figure 1-4, multiple new technologies, each potentially transformative, are now upon business decision makers. But have I dwelt too much on IT? I don't think so, because as you will see in subsequent chapters, the IT infrastructure of an organization is so very key to high-performance data-to-learning-to-action processes. Yes, people and process are undoubtedly even more important, but there is no getting around the fact that as technology progresses it becomes an ever-increasing part of the performance-improvement equation. Especially so with the ever-increasing impact of machine-based learning.

Have I dwelt too much on the confusion aspect of disruptive technologies at the expense of all the upside these technologies will certainly ultimately deliver? Perhaps I'm guilty of that, but the concern is, again, that confusion results in either paralysis or recklessness if there is not a proper framework applied to put order to the confusion. And, of course, there is no denying the reality of the economic data, such as the deteriorating trend in ROA, that bolsters the concerns.

Furthermore, the complexity and confusion that come with the multiple, parallel, and rapid advances of technology are amplified not only by the combinatorial effect just outlined, but also by the intense competition of all the various players that seek to supply these technologies. The reality is that the dynamics among competitors and the resulting competitive outcomes are difficult to predict, particularly because there are often game theoretic aspects to the competition with no long-term stable equilibria. It is a game of coopetition, in which players sometimes cooperate—on, for example, standards—but compete in other areas. And competitive and cooperative postures can quickly switch places. Further, as we all know, what is said for public consumption and for marketing purposes may not precisely reflect reality. Today's technology marketplace represents a process of creative destruction that would no doubt awe Schumpeter.

So, making decisions about technology in today's environment is tough. But not making decisions is not a viable option, nor is merely "going along for the ride with suppliers" a viable option. Ensuring that the best decisions are made with respect to information technology is a significant aspect of what the optimization of data-to-learning-to-action processes is meant to address. When others are confused, that presents opportunities for you. Don't let *your* organization be the Dilbert cartoon!

Why Have Legacy Approaches Come up Short?

The fact that business performance has been lagging for quite some time is not for the lack of organizations that are trying to improve. A quick look in the business section of any airport bookstore or the more extensive selections of Amazon reveals myriad prescriptions for better business performance; for example, perspectives on business strategy, operational excellence, quality improvement, organizational design, leadership, and so on. Most of these perspectives are very valuable, and, by all means, they should be put into practice. But they clearly haven't been *sufficient*. Why not?

Frankly, because they generally do not directly address the overarching processes that are at the core of all businesses: *learning*, and the making of decisions based on the learning. Yes, there are certainly valuable perspectives on organizational learning. But they are often quite generalized and not necessarily in the context of specific types of decisions in the specific environments that matter most. They also typically fail to define learning robustly enough and rigorously enough to enable both the generalization and the quantification that is required for pervasive and truly measurable applications. There are also sound perspectives on decision science and analysis. But the field of decision analysis has been underutilized, perhaps in part because it has historically been oriented toward very large, infrequent decisions rather than the types of decisions that constitute the bulk of the decisions that are made daily in organizations and that therefore provide the greatest share of leverage for performance improvement. There are excellent perspectives on process improvement. But historically business-process improvement has been oriented toward the transaction-based processes of the organization rather than the learning and decision-intense processes that are increasingly the core drivers of value for most businesses.

What has been missing is an integrated and rigorous perspective on improving that universal process—data-to-learning-to-decision-to-action—which underpins all the other processes, activities, and directions of an organization. In the next chapter, we will delve into how this missing perspective can be adapted from the antecedent perspectives highlighted here and integrated into a method that has the power and flexibility to help today's businesses achieve better results in our current dynamic and difficult environment.

The Case for Data-to-Learning-to-Action

Given the current business reality that we have summarized thus far, characterized by declining returns on assets and high topple rates, buffeted by wave after wave of exciting but also disruptive technologies, increasingly awash in data, and increasingly a world in which *learning* is no longer a term relevant only to humans, but also to machines, the time has clearly come to seek out new ways for achieving sustainable performance improvement that has eluded legacy approaches.

What this book is about is demonstrating that optimizing data-to-learning-to-action is *the* paradigm and associated method that can ensure continuous performance improvement for today's businesses. That it can do for the core of contemporary organizations, knowledge and learning-based activities and processes, what the reengineering revolution did for transactional-based processes a decade or two ago.

The optimizing data-to-learning-to-action approach is geared to holistically optimize the chain of activities that span from data to learning to decisions to actions, an imperative for achieving outstanding performance in the contemporary dynamic and competitively intense business environment. Adapting and integrating insights from decision science, constraint theory, and process improvement, optimizing data-to-learning-to-action is a method that is clear and effective and can be applied to nearly every business function and sector. It is characterized by the technique of systematically working backward from decisions to data, estimating the value of the learning flow along the chain, and identifying the inevitable value bottlenecks. Moreover, it provides techniques for quantifying the value that can be attained by successfully addressing the bottlenecks, providing the credible support that executives need to make the right level of investments in the right place and at just the right time, as well as providing enhanced support for all those other day-to-day decisions that occur within an organization.

Importantly, in today's dynamic technology environment, with its never-ending stream of new technology options that decision makers must continuously consider and reconsider, the optimizing data-to-learning-to-action method provides the means for making consistently effective decisions about these technologies for specific organizational environments, underpinned by overall business strategy and credibly quantified value. And while the dynamism of the technology marketplace tends to grab the headlines, optimizing data-to-learning-to-action is about much more than just technology. It is about holistically optimizing people, process, and technology aspects across the data-to-learning-to-action chain, and doing so on a continuing basis.

In summary, the optimizing data-to-learning-to-action method presented in this book will deliver the following for your organization:

- A clear and effective method that can be broadly applied to ensure continuous performance improvement for organizations operating in today's challenging environment

- A means to identify, quantify, and resolve performance-limiting bottlenecks in data-to-learning-to-action processes, the key to overall business performance

- A way for getting the absolute most out of people, process, and technology investments, and doing so on a continuing basis

Summary

To summarize the key points of this chapter, we reviewed data that suggests that the fundamental economic performance of businesses has been deteriorating. At the same time, unprecedentedly rapid advances in technology in multiple fields, particularly information technology, promise significant upside but also engender a decisioning environment characterized by increased complexity and confusion. It is clear from the economic data that the traditional toolkit of business-performance improvement currently fails to adequately address the realities of today's business environment, and has actually failed to do so for some time. Optimizing data-to-learning-to-action is designed to do what the traditional toolkit does not—meet the challenge of delivering continuous business-performance improvement for today's businesses.

Roots of a New Approach

New ideas are invariably rooted in prior concepts—very often combinations of prior concepts. And for synergistic combinations to be developed, the prior concepts often first need to be generalized or extended so that they can be flexibly recombined into something for which the whole is greater than the sum of the parts. So it is with the optimizing data-to-learning-to-action approach. In its case, it is primarily rooted in three fields of management science:

- Process improvement
- Theory of constraints
- Decision science

And in each of these fields, the path to our new approach is to first generalize the key concepts and then combine those concepts in a synergistic way.

Root 1: Process Improvement

A revolution in thinking about improving business performance that centered around *processes* emerged in the 1980s and came to full fruition during the 1990s. This emphasis on processes had its roots in the quality-improvement movement, which was a reaction, particularly in the United States, to the

© Steven Flinn 2018
S. Flinn, *Optimizing Data-to-Learning-to-Action*,
https://doi.org/10.1007/978-1-4842-3531-7_2

perceived higher quality of products that were being manufactured in Japan compared to those made in the United States. Although it may now seem like an over-reaction with the benefit of hindsight, beyond just product-quality advantages, there was a significant concern in that era that Japan was poised to overtake the United States economically more generally. This perceived external threat served to significantly amplify the motivation for improving the performance of Western businesses and redesigning processes became a leading approach that was applied to achieve the required performance improvements.

The process revolution accelerated in the 1990s, in good part because of its popularization by Michael Hammer, as well as his expansion of its application areas under the overall banner of "reengineering."[1] This had a significant impact on large organizations, with many having one or more process-improvement or reengineering initiatives ongoing at any given time during the 1990s and well into the next century. While never completely dying out, reengineering settled into more of a slow burn over the past decade, perhaps partly because it felt a little bit like old news, and perhaps partly because it had reached a level of saturation in its natural target areas for many organizations.

On this last point, I will have a good deal more to say throughout the book, but it should be noted that process improvement historically has predominantly been focused on the more transactional parts of the organization; for example, manufacturing, procurement, and customer service. Rarely was there a focus on, for example, research and development. This bias was exemplified by the predominant types of information technology that were applied in process-reengineering initiatives, enterprise resource planning (ERP) systems, which were at least initially primarily focused on supply chain–related functions. There was really no compelling reason for the bias toward more transactional functions from the standpoint of the process paradigm itself. It was seemingly more of a tradition, again perhaps stemming from its linkages to quality improvement, which was naturally more associated with manufacturing and the transactional-based functions of the organization. Hammer and others did little to change that focus, although Hammer did attempt to extend reengineering to management in general. But at that point there had also been a good deal of dilution from the original process redesign concept.

---

[1] Hammer, Michael, and James Champy, *Reengineering the Corporation: A Manifesto for Business Revolution* (HarperBusiness, 1993).

So, why was there limited interest and application of process improvement in functional areas such as R&D? I offer some hypotheses, categorized under the themes of "complexity," "learning and adaptation," and "everyone but me," as follows:

- **Complexity.** In applying any new approach, the simpler areas of application are naturally addressed first. It is relatively easy to map out the typically quite linear process steps for the more transactional parts of an organization. It can be much more difficult to do so for the complex and non-linear steps that are characteristic of the more "messy" functional areas in an organization. In functional areas in which the words *creativity*, *innovation*, and just plain *learning* are commonplace, for example, it is a good bet that process steps will not be simple.

- **Learning and adaptation.** Process improvement informed by the reengineering paradigm traditionally struggled with learning, and adaptation most generally. It more comfortably handled situations in which a process was redesigned and the new design was then semi-permanently established. Certainly, this traditional process redesign accommodated learning by individual actors in the process, but it didn't really directly address continuous learning occurring throughout the process, with that learning then becoming institutionalized within the process.

- **Everyone but me.** There is a natural tendency for people, including business executives, to be more favorably disposed to other people undergoing a change rather than making a change themselves. So, when performance-improvement targets are identified, they naturally tend to be in those areas of an organization that are not seated closest to the power centers, which means that the identified areas are less likely to be truly core to the company, the truly core areas being the more decision-rich areas of the company and those areas particularly rich in the types of decisions that have the greatest financial impact.

So, in summary, the process paradigm failed to "take" in many parts of the organization, and mostly for no good reasons at all! However, the struggle of classic process redesign to comfortably accommodate learning was a more fundamental issue, and that is something we will explicitly address when constructing our new approach.

Let's recall the basics of processes. A process is simply a series of activities that leads to an outcome, thereby representing the fundamental structure of businesses, even if that structure is not necessarily explicitly identified on a chart somewhere. Adam Smith's famous illustration of the economy at work, his pin factory, was quintessentially a description of a production process. It follows that to improve business performance, it is *necessary* to address the business's processes. Doing so may not be *sufficient* for achieving a business-performance improvement, which is why other performance-improvement methods often need be applied, but improving processes is surely necessary.

The fundamental law of business processes is that *all* business functions can be decomposed into process steps. It is just that some are much more complex than others, often with more feedback loops contributing to the complexity. Fundamentally, all processes comprise one or more *actions* (i.e., activities). And fundamentally, all processes in which intelligent actors such as people perform actions comprise one or more *decisions* (if intelligent actors are being wasted by being applied in processes that require no decision making, there is something clearly wrong!). And these decisions by intelligent actors are made on the basis of *data* or information. Therefore, all people-based processes are composed of a series of data-to-decision-to-action steps or sub-processes. This simple insight comes in handy because it enables us to address even the most complex process areas of an organization.

But there is something missing here—the *learning* part of our data-to-*learning*-to-decision-to-action chain. That's the part that was not adequately treated in classic process improvement, but we will do so here, while also putting the overall concept of "learning" on firmer ground conceptually. When we do so we will find that our universal process has universal performance-improvement applications throughout an organization.

Root II: Theory of Constraints

The theory of constraints, while there were precursors, was primarily developed by Eliyahu Goldratt, who published the popular management book on constraint theory in story form, *The Goal*, in 1984, and followed up with a series of additional books on the topic.[2] The core concept of the theory of constraints is that achieving a desired outcome or goal is inevitably governed by a limiting constraint or bottleneck. Specifically, the constraining factor limits

---

[2]Goldratt, Eliyahu, and Jeff Cox, *The Goal: A Process of Ongoing Improvement* (North River, 1984).

the *throughput* of some flow on which the goal is dependent. For example, the flow of assembled parts in a manufacturing environment, or money generated by sales. Most generally, the throughput of a system is a *rate* metric, the flow or production of goal-related units per a temporal metric, such as assembled widgets per hour.

The fundamental law of the theory of constraints is that there is *always* a limiting factor in any system, which implies that resolving one constraint always leads to another constraint! So, where does it end? It ends when the marginal cost of resolving a constraint exceeds the benefit of the increase in throughput that resolving the constraint would have on the system's overall goal. As in the case of all economic-based decisions, we keep taking actions until the marginal cost of the next action exceeds the marginal benefit of that action.

How are constraints resolved? That depends on the exact nature of the system and generally first requires some solid analysis to understand the overall dynamics of the system. For example, if the system has a very dependable flow between the stages of a process, then perhaps the limiting constraint could be addressed by simply adding capacity at the bottleneck; for example, by employing two machines in parallel rather than just the one original machine. However, in many systems there may be significant variation of flows resulting from any number of factors. In such cases, inventory—or, more generally, buffers—may be the more effective way to address the constraint than by adding capacity. Again, getting to the right answer of where exactly the true limiting factor is and what should be done about it can require significant analysis for complex, stochastic-based systems.

Now, the theory of constraints is a very broadly applicable concept. Goldratt and his precursors, at least initially, tended to focus on manufacturing processes, for which the approach is quite intuitive. But it can (and should!) be applied to any type of process. In our case, we want to apply it to our universal process, the data-to-learning-to-action process. We want to be able to identify the bottlenecks in that process and then determine the best way to resolve them.

But there's just one problem. To do that, we need a way to understand what a bottleneck in a data-to-learning-to-action process is costing us, which in turn will tell us what the value would be of resolving the bottleneck. But to understand the cost of the bottleneck we need to understand what exactly the throughput of the data-to-learning-to-action process is and what the value would be for various levels of that throughput. And for that we need something more than just the process paradigm and the theory of constraints. We need to turn to decision science.

Root III: Decision Science

The field of decision analysis (which can be considered a major sub-field of the decision sciences in general) was in good measure pioneered by Ronald Howard of Stanford University, whom the author was fortunate enough to have as a professor at Stanford's Engineering-Economic Systems department.[3] The field was developed to help decision makers make decisions in a consistent, rational way. It is therefore a prescriptive field of decision science—i.e., relating to the way decisions *should be made* (i.e., normative) as opposed to the study of the way human decisions are *actually* made (i.e., descriptive). And human decision makers need this prescriptive help because of two factors that tend to be problematic for people, particularly when making business-related decisions: complexity and uncertainty.

The first issue is intuitively clear—for big, complex decisions in which there are a lot of moving parts or intermediate choices, people need a structured way to think about the decision or sequence of decisions. And, of course, many business decisions fall into this category. To try to tackle a complex decision by merely having free-form discussions with colleagues is as likely to be effective as performing well in a chess game without understanding how chess openings are structured, what end-game positions are known to lead to what outcomes, and so on. In this regard, decision analysis is similar to our other foundational roots, the process and constraint paradigms, in that they are all are structured methods that help people achieve desired objectives in complex environments.

The second factor that is problematic for human decision making, uncertainty, is at the core of the decision sciences. Of course, our typical use of the word *decision* is predicated on the existence of some uncertainty. If there is no uncertainty, then it hardly qualifies as a decision in our usual, human, sense of the word. There can certainly be *actions* without uncertainty, but the performed actions are pre-determined, such as, for example, in the manner of the simple automation of a computer program: "if condition x occurs, then perform action Y."

People have an array of well-known biases and inconsistencies when it comes to probabilistic reasoning.[4] Short-hand decision-making heuristics evolved that were presumably optimal for our ancestral hunter-gatherer environments, but not so much for complex business decisions. Decision analysis provides a logically consistent structure that can be applied to address decisions that are made in the face of uncertainty.

---

[3]For a comprehensive review of decision analysis, see Howard, Ronald, and Ali Abbas, *Foundations of Decision Analysis* (Pearson Education Ltd, 2016).

[4]See, for example, Bang, Dan, and Chris Frith, "Making better decisions in groups", *Royal Society Open Science*, August 2017. http://rsos.royalsocietypublishing.org/content/4/8/170193

A decision analysis can be quite a significant undertaking, and for large decisions, that undertaking and expense can be quite warranted. But because of this significant investment in time, expense, and expertise, decision analysis is typically applied only to very large decisions; for example, whether to invest in a new manufacturing facility, whether to acquire another company, or whether to launch a new product line. The decision analyses are also often treated as "one-off" projects, performing only an analysis of a specific decision while taking as a given the organization's overall people, process, and technology infrastructure for supporting decisions in general.

But what about all those other decisions that are made daily in an organization, and particularly those ongoing decisions with respect to that people-, process-, and technology-based decision-support infrastructure? If they are not nearly optimal, that clearly represents a huge loss in value. And yet, it may simply not be cost and/or time effective to apply classic decision analysis to all these decisions. Is there another way that leads to better decisions generally, but does so cost effectively? Yes, but we will need to creatively combine concepts from decision analysis with our other roots to do that.

A key concept of decision analysis is "value of information." The concept is based on the fact—though a not necessarily intuitive fact—that information only has value to the extent that it can effect a change in a decision. In other words, if it is already a foregone conclusion what a decision is going to be regardless of an additional set of information, then that additional information is not worth anything. This seems pretty clear when you think carefully about it, but it is amazing how often organizations spend money on gathering additional information when the decision is already pre-determined!

Admittedly though, things are not quite this black and white, because the reality is that information has no value at all if it will not *potentially* change a decision. It is the "potentially" aspect that makes things tricky, because it can apply to possibly affecting a currently known decision, but it can also apply to actions associated with a decision, or even a type of decision, that has not currently been determined. So, we must be a little careful in applying this thinking, and not apply it so narrowly that we exclude deriving insights from information that may lead to entirely new areas of value and associated decisions.

If additional information can potentially change a decision, it does have value. And that value can, at least in principle, be calculated by comparing the value of the action or actions that are expected to result from the original decision with the expected value of the action or actions resulting from a different decision that is made with the benefit of the additional information.

Information that can potentially change a decision has tangible value.

The value of information concept can therefore inform the sub-decision of whether it is worth gathering additional information. And, specifically, as is again always the case in economic-based decisions, it is worth gathering additional information if the marginal value of the additional information exceeds the marginal cost of gathering the information.

Additional information can affect a decision because the information contributes toward reducing the uncertainty that the decision is dependent upon. That uncertainty can in some cases completely vanish based on the new information, but more typically the new information serves to reduce uncertainty but does not completely eliminate it. Medical diagnostic tests are obvious examples—they tend to be useful, but are certainly not fool-proof, typically yielding some degree of both false positives and false negatives. But that doesn't mean they are not useful! It is simply that they are not as valuable as they would be if they were even more accurate.

For general purposes, and particularly for our purpose here, the concept of value of information, while very useful, is a misnomer. Raw information is not what has value—it is the translation of that information into revised views on uncertainties, or, more formally, into changes in probabilities, that has value. In classic decision-analysis projects, that translation is explicitly performed within various types of models, specially selected and tailored to be appropriate for the decision at hand. Outside of those special projects, in the general business environment, the translation of information into adjusted views of uncertainties is typically performed in a more informal manner by one or more people, assisted by a variety of information technologies. We already have a common term for this general type of translation process: *learning*.

Fundamentally then, learning can be thought of as the process of translating information into reduced uncertainty, thereby improving the ability to predict. That is true for people, for any organism with neurons, as well as, increasingly, for machines. If we want to be even more precise, any given specific instance of learning is associated with an adjustment to one or more probability distributions that serve to embody uncertainties. Typically, this would result in the narrowing of probability distributions, which is aligned with having tighter bounds on a working hypothesis of potential outcomes, but it can also be the case that there is a broadening of probability distributions if the new information that is attained is disconfirming of a current working hypothesis. It is well known from the field of cognitive science that people have a bias toward assimilating confirming information with respect to their current hypotheses about the world rather than seeking out and/or assimilating disconfirming information, and therefore disconfirming information should be actively sought. Nevertheless, as a good approximation over the long term, and as illustrated in Figure 2-1, the essence of learning is the *shrinking* of probability distributions.[5]

Learning is the process of reducing uncertainty.

Perhaps that perspective sounds somewhat constraining, but think about it— why else do we learn? We want to know more about something that we don't already know about or don't understand. Our uncertainty of what that subject is about or how it works becomes more constrained after we learn. We are more certain than we were before, which often translates into an increased feeling of confidence about the subject.

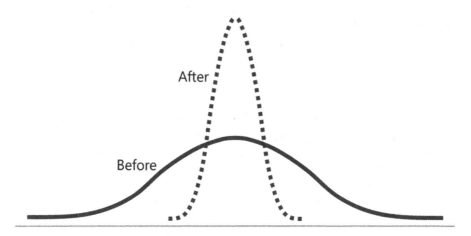

Figure 2-1. Learning is a shrinking of probability distributions

Now, for our personal learning, that reduction in uncertainty doesn't necessarily have to potentially change decisions—we may want to learn about something for the pure pleasure of learning, and therefore we may not ascribe a value to the learning. But in the business context, reductions in uncertainty that affect decisions are valuable, and hence learning most definitely in that context has a tangible value.

---

[5]Figure 2-1 more formally represents a probability *density* function. In this book, however, we simply drop the term *cumulative* from the term *cumulative probability distribution* when we want to distinguish a probability density or mass function from its associated cumulative probability distribution.

Notice also that because learning can be defined as a process for making *changes* to probability distributions, it immediately follows that learning must necessarily be defined *over time* rather than as a *point in time*, which is very different from, say, information or knowledge, which can be in theory measured at a given time. Therefore, there is a sense that, unlike data, information, or even knowledge, learning represents a *flow*.

The Emergence of the New Approach

So, the field of decision analysis provides the concepts that enable us to put the idea of learning on a solid foundation that, among other benefits, can support quantifying the value of learning. And from the process paradigm, we know that the data-to-learning-to-action process must have a flow. What is that flow? It is learning. Specifically, learning that is actionable; that is, learning that has the potential to change a decision. Which means learning that has tangible, quantifiable value.

Actionable learning has tangible, quantifiable value.

The theory of constraints provides a powerful perspective related to learning and our universal process as well: if learning is the flow of the data-to-learning-to-action process, the quantity of actionable learning per unit of time can be considered a *throughput* of the process. That is, what we are trying to optimize with respect to the data-to-learning-to-action chain is the *rate of learning*. Again, not just any learning, but *bona fide* quantifiably valuable learning, or what we can term in short-hand as *actionable learning*.

This flow of actionable learning along the data-to-learning-to-action process comprises the continuous conversion of a forward-feeding stream of information, knowledge, and associated predictive models, whether in minds or machines, into reduced uncertainty. Whether we refer to learning as reducing uncertainty, or as flow of the data-to-learning-to-action process, it is referring to the same phenomenon. It is just that learning flow represents an aggregated perspective of reducing uncertainty across multiple sequential— and in some cases, parallel—steps of the process.

We don't necessarily see probabilities shrinking in the specific form that is depicted in Figure 2-1 throughout this flow. There are many different representations and structures that can embody the reduction of uncertainty. For example, if we examine the human brain, we see nothing that looks like Figure 2-1, only an immense network structure composed of neurons and synaptic connections. And yet the human brain is capable of probabilistic assessments. In fact, we now have a good idea what region of the brain enables our sense of differing levels of uncertainty and changes to the level of uncertainty.[6] Similarly, in an organization, learning and its results—updated views on uncertainties—can be embodied in many different forms, forms that do not necessarily look anything like Figure 2-1. They may be embodied in computer-based forms such as conditioned data sets, documents, spreadsheet models, and so forth, as well as in the minds of the members of the organization. Only rarely (but more often, if this book has any say in it!) are uncertainties represented by forms such as that of Figure 2-1.

The theory of constraints tells us something else: there is inevitably a learning constraint or bottleneck in a data-to-learning-to-action process, just as is the case for any other type of throughput in any other type of process. We can therefore improve the data-to-learning-to-action process by identifying the primary constraint, resolving it, determining the next constraint, resolving that one, and so on. The process continues until when? Yes, until the value of resolving the constraint no longer exceeds the cost of doing so.

There always exists a learning constraint in a data-to-learning-to-action process.

It is readily seen then, that unless we can at least make rough estimates of the value of learning there is no way we can effectively perform this logically iterative approach. Without such estimates we essentially fly blind, not understanding the true costs of bottlenecks, and therefore either dither or do nothing, or throw away time and money on the wrong things. In other words, very often, the current state!

One important difference for our data-to-learning-to-action process compared to a classic process of producing widgets is that our universal process "chain" can be a loop. That's because, as illustrated in Figure 2-2, an action that occurs as a result of the process may well create new data that, in turn, serves as an input into learning, which affects decisions that lead to new actions, and so on. I will often refer to the data-to-learning-to-action process as a "chain" in the book, but it should be understood that the process can really be a chain plus a feedback loop. In fact, some actions may be performed solely to generate additional data that can be beneficially learned from. We have a common name for those types of actions: *experiments*.

---

[6]"Two Distinct Brain Regions Have Independent Influence on Decision-Making", *Neuroscience News*, September 2017, http://neurosciencenews.com/decision-making-brain-regions-7390/.

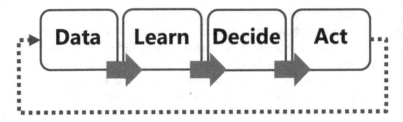

Figure 2-2. The data-to-earning-to-action loop

Another important aspect of data-to-learning-to-action processes is that they may be *nested.* That is, a data-to-learning-to-action process may include subsidiary data-to-learning-to-action processes within it. For example, as we will see, determining how to address a bottleneck in a data-to-learning-to-action chain is itself a data-to-learning-to-action chain.

So, data-to-learning-to-action processes can be both recursively closed-looped as well as nested, which makes for some significant complexity! This is why we need a structured but flexible approach to be able to continuously optimize such processes, while also always remaining mindful that applying estimates and approximations is often more than sufficient to significantly improve over the current state.

Summary

In summary, optimizing data-to-learning-to-action has its roots in the management-science fields of business processes, constraint theory, and decision analysis. Data-to-learning-to-action is a process. In fact, it is the universal process, and because it is a process it has a flow. That flow is actionable learning, which is the conversion of a stream of information, knowledge, and associated models, whether in minds or machines, into reduced uncertainty. This actionable learning is learning that can potentially affect decisions, which means that it has quantifiable value. Constraint theory tells us there is always a limiting constraint to any flow, which means there is always a limiting factor with respect to actionable learning. With these fundamental concepts in place, we are now ready to move on to all the details of optimizing data-to-learning-to-action.

Data-to-Learning-to-Action

Now that we have defined the basics of the data-to-learning-to-action process, it's time to dive into the details of its intermediate steps. And while birds do it, we do it, and even educated machines can now do it, our focus, of course, will be on the elements of this universal process that specifically pertain to its execution within organizations.

Elements of the Chain

Figure 3-1 depicts the major elements of data-to-learning-to-action chains operating within organizations. I say "chains" because there will be many different data-to-learning-to-action chains in an organization. For example, HR-related data-to-learning-to-action chains, marketing-related chains, R&D-related chains, and so on. While the elements depicted in Figure 3-1 are universally applicable, it should be acknowledged that the emphasis on the individual elements can be different for various application and functional areas. Certainly, it must also be noted that there can be a degree of fuzziness and overlap with respect to the labels that have been chosen for the elements of the chain. One could argue for different nomenclature here and there, and for some applications, additional, finer-grained elements may be appropriate.

© Steven Flinn 2018
S. Flinn, *Optimizing Data-to-Learning-to-Action*,
https://doi.org/10.1007/978-1-4842-3531-7_3

Figure 3-1. Elements of the data-to-learning-to-action chain

However, as we walk through the elements, I think you will come to see that they can quite effectively model any data-to-learning-to-action chain, at least on a first-cut basis, enabling good clarity on what activity of a real-world data-to-learning-to-action process maps to what element of Figure 3-1. In fact, by viewing each of the elements in the context of its specific role within the whole of the process chain rather than merely as an isolated concept, clearer thinking can be promoted on the essential nature of the element.

We're going to tour each of these elements, but first, you may be wondering, "Where did the 'learning' go from our basic data-to-learning-to-action loop from the last chapter?" The learning is still there—it simply extends across the chain, comprising a forward-feeding flow of uncertainty reduction all along the data-to-learning-to-action chain, as represented by the arrows.

Learning can be analogized to other phenomena that are fundamentally the same but that seem to have two different natures. For example, in physics: Is it a particle? Yes. Is it a wave? Yes. Or, from neuroscience: Is it a cluster of firing neurons? Yes. Is it a mental state? Yes. So it is with learning. Is it the process of reducing uncertainty? Fundamentally, yes. Is it a flow? Also, yes. In later chapters, we will see how these two natures of learning can be alternatively represented; for example, by the flow form that is exemplified by Figure 3-1 or by an uncertainty-based form that we call a *learning-value diagram*.

In recognition of a more traditional, narrower notion of learning, Figure 3-2 depicts roughly how the standard views of learning and managing information and knowledge map to the chain, with learning mapping to the "downstream" elements and managing information and knowledge mapping to the "upstream" elements. But as we'll see later in this chapter, consistent with the perspective of learning flowing forward all along the chain, and as is represented by the forward-flowing arrows in Figure 3-2, learning is increasingly being embedded directly into what would traditionally be considered the lower-level information processes and structures of the upstream elements of the chain.

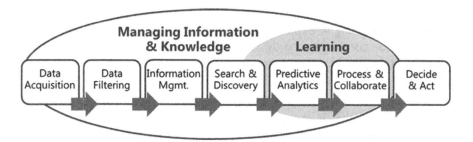

Figure 3-2. Traditional notion of learning mapped to the data-to-learning-to-action chain

We'll work our way forward along the flow of learning of the chain in this chapter, starting with the Data Acquisition element. But before we do, an important point needs to be reinforced about the elements of the chain in general: *the elements of the data-to-learning-to-action process can apply to humans, to computer systems, or to a combination of both.* To attempt to distinguish any of the elements based upon whether they are automated or not, or even to simply generally assume, for example, that when the terms *knowledge* or *learning* are used that these references must be to activities or capabilities that are the province of humans, or that the term *data acquisition* firmly connotes computer-based activity, is "old-think" that can stand in the way of optimizing contemporary business performance. Such old-think simply doesn't comport with current perspectives of the field of cognitive science or with the inevitable trends in the world of technology, which in many ways are recapitulating biological-based processes. So, in the following tour and subsequent chapters, I will often discuss the elements of the data-to-learning-to-action process with respect to both minds and machines, as well as an integrated system of people and systems.

Data Acquisition

Data refers to raw, unprocessed information and, of course, comes in many different forms. It may, for example, comprise measurements of a physical phenomenon, it may be in the form of text, it may be in the form of speech, it may be images, or it may be numerical in nature. It may constitute behavioral information associated with people. The acquisition of data may be performed by instrumentation such as sensors or cameras, through standard computing devices with keyboards and/or microphones, or directly by people using their various senses. Data may be generated as part of an intentional data-gathering process or may be a by-product of other activities.

Of course, as we have discussed, the quantity of data that is available for subsequent processing is exploding, resulting in "Big Data." The internet of things is a major contributor to this explosion of data, and that phenomenon has only just begun in earnest, so we are destined to have continuously "bigger" data.

On the other hand, it is still quite often the case, even with today's surfeit of data, that the data that is most required for some purpose is not available, or is not available in sufficient quantities. It may be that it is simply not being captured at all because its relevance wasn't previously recognized, or that it exists but has never been available for the intended application, or that the quality of the currently available data is too poor for the intended application.

So, it is often necessary to specifically identify missing data that would be valuable to the overall data-to-learning-to-action process and determine a means to acquire it. That can require some creative thinking. And, as we will see in subsequent chapters, data that can serve as a basis for improving predictions about uncertain factors that influence decisions is data that is valuable, and the better the data enables us to make *perfect* predictions on the outcomes of actions, the more valuable the data generally is. The quest for better predictability implies that even with ever-increasing volumes of data available, it will never be enough—there will always be data that we don't currently have that would be valuable to us and that we will therefore continuously strive to attain.

Data Filtering

In many cases, data must be filtered to make it fit for subsequent processing. Our brains perform this filtering all the time, for example. The information flooding in from our senses passes through a series of filters so that our conscious experience is not constantly overwhelmed by less-relevant information. Of course, even these exquisitely evolved functions for determining signal from background noise can find themselves challenged in the modern workplace with its hundreds of urgent daily emails!

Data often needs to be conditioned for application in the downstream elements of the data-to-learning-to-action process. Spurious or outlier data may need to be removed, for example. This is often a key application area of data science and an activity of data scientists, who require data to be of sufficient quality to serve as inputs into subsequent steps that are aimed at deriving predictive insights from the data. It is often said that more data beats better algorithms in deriving predictive insights. While not necessarily always the case, there is certainly a nugget of truth to this data-science shibboleth. But attaining more data quite typically means attaining data that is of more marginal quality than that which has already been attained, and so this additional volume of data will require extra conditioning to increase its quality sufficiently to fully extract its potential.

A caveat about data and data science in general: if you have a discussion or read an article in which the word *data* is used numerous times but there is no mention of the word *decision*, or the ratio of the word *data* to *decision* is enormous, that is a red flag signifying that the tail may be wagging the dog. And, more to the point, it may signify that there is not an appropriate understanding of the *value* or lack thereof of the data that is being discussed, rendering informed decisions on investments associated with acquiring and processing the data impossible.

Dilbert cartoon alert: using the word *data* without also using the word *decision!*

Most generally, the data-filtering step of the data-to-learning-to-action process is about identifying and removing "noise" from the data, just as our brains continuously do. We may not yet be at the stage of determining "signal," which typically comes later in the data-to-learning-to-action process, but data filtering is the first step in the overall process of detecting signal or insights from the background noise embodied by the data. It can therefore also be considered an initial stage of the flow of learning of the process, reducing uncertainty by eliminating, or at least reducing, noise from data sets. In recognition of the data filtering and transformations performed by this element, we typically apply the term *information* to the output of the element to distinguish it from its input—unconditioned, raw data.

Information Management

The filtered data, or information, is organized appropriately and is typically persistently stored for use in subsequent steps of the data-to-learning-to-action process, which is the province of the Information Management element of the process. The capabilities included in this element have been evolving rapidly, with a particularly important trend being an increasingly tight binding of information structures and learning (an overall architecture that is more in line with the way our brains do it), and so we will spend a bit more time on this element during our tour.

The Information Management element includes overall information organizing and governance considerations, including decisions and associated implementations with respect to where to physically store the information, how to organize the information and the associated structures that are applied to effect the desired organization, as well as requirements for pre-processing of the information prior to storage. This is also the element of the data-to-learning-to-action process that includes basic security and privacy considerations, such as who, and what systems, have access to what information. These security and privacy considerations inform decisions on whether the information is stored in the cloud or on-premises within private firewalls. Legal discovery issues may also need to be considered in determining how information is stored and who has access to the information.

Organizing and structuring information most fundamentally involves forms of categorization and classification, including considering and implementing multiple *levels of abstraction*. For example, at a higher level of abstraction, formal *taxonomies* and informal *folksonomies*, or, even more generally, *ontologies*, may inform how the stored information is organized, potentially requiring the application of taxonomy and ontology expertise to determine the best structures to fit the anticipated use cases. These ontological structures can be specific to an organization, and increasingly they comprise general-purpose knowledge bases that can provide some degree of common sense or *semantic* understanding about the world, and impart that understanding to applications associated with subsequent elements of the data-to-learning-to-action chain.

At a lower level of abstraction, alternative database architectures such as relational or "NoSQL" databases or structures may be considered by database architects, guided by the top-down requirements of the higher level of abstractions and the anticipated use cases. Other information structures may be implemented for special purposes; for example, data warehouses that are specially structured to support analytic-based use cases, and search indexes. The choices of specific information structures may be influenced by whether the information is primarily going to be directly accessed by people or by systems, as well as by the anticipated types of searching, discovery, and processing that will be performed against the information.

The Architecture of Learning

The trend toward binding learning capabilities and the resulting inferences within information structures is revolutionizing enterprise computing, which, again, is in many ways recapitulating with technology the way our brains work. The core structure underpinning this revolution is what I termed in *The Learning Layer* "the architecture of learning" and is illustrated in Figure 3-3. This structure exemplifies the general trend of learning capabilities' increasingly becoming directly embedded in what would historically be considered lower-level structures. In other words, that shaded area of Figure 3-2 representing a traditional view of learning is inevitably creeping backward along the chain, reinforcing the optimizing data-to-learning-to-action perspective that learning is best considered as flowing across the *entire* chain.

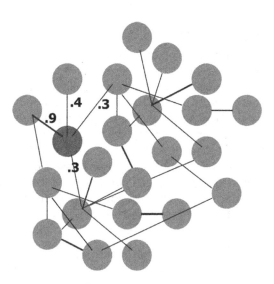

Figure 3-3. The architecture of learning

The architecture of learning is a network-based structure comprising nodes, whereby the nodes may be, for example, objects that comprise content or that reference items of content, or that represent users of the network. The network further comprises weighted relationships among the nodes. The weighted relationships may be directional in nature. This is the basic structure of the human brain, of artificial neural networks, and, more generally, of system-based structures that can encode learning by adding nodes and/or changing the weights of the relationships among the nodes.

While the processes for adjusting these relationship weights—which constitute the actual process of learning—can be quite different for biological-based neural networks, artificial neural networks, and the more basic network architectures that support general enterprise machine-learning applications, the underlying network structure that enables such learning processes is similar.

The architecture of learning can be considered a "fuzzy network," reflecting the fact that the relationships among the nodes are by degree rather than merely binary relationships.[1] Or, more popularly now, in accordance with the mathematical field of graph theory, the architecture of learning of Figure 3-3 is typically described as a "graph," with the relationships among the nodes being termed "edges."

---

[1]Adaptive fuzzy networks and their applications are described in detail in Flinn, Steven, and Naomi Moneypenny, "Adaptive Recombinant Systems". *World International Property Organization*, publication no. WO/2005/054982 (June 16, 2005).

Anticipatory Computing

The integration of a network of objects representing people, along with objects that reference content or topical areas, all within a common network paradigm, is a key prerequisite for a core capability of modern computing systems and an accelerator of human learning, *anticipatory computing*— system functionality that can anticipate what a user will find relevant without necessitating being directly instructed by the user. The other prerequisite for anticipatory computing is that the values associated with at least some of the connections or edges of the graph are adjusted based on the processing of usage behaviors of individual users. Such continuous adjustments constitute a form of learning that is, again, loosely analogous to the adjusting of connections among neurons in our brain in response to the external information that we process.

A familiar example of such networks is the *social graph*, which describes the connections among people in a social network. The term was originally introduced by Facebook, but it applies to any system in which people connect with one another and whereby those connections may be symmetric (e.g., "friending") or asymmetric (e.g., "following"). Even in its original incarnation in which its nodes only represent people and not also content, useful insights and anticipatory inferences can be derived from the social graph by the downstream elements of the data-to-learning-to-action process, Search and Discovery, and Predictive Analytics.

For example, common connections among two or more people can be identified, and there is the ability to calculate the degree of separation between any two people. Although more sophisticated network characteristics can be derived from the social graph, even these simple constructs can be quite usefully applied. For example, a system can leverage the social graph to suggest to you that you connect with someone who is connected to one or more people in common with you. This is the basic approach behind the people suggestions you receive on Facebook, LinkedIn, and Twitter, as well as with many enterprise social platforms, and is a simple example of anticipatory computing.

More inferential and anticipatory power can be derived from graphs that *integrate people and content* and that can thus capture user behaviors with respect to one another and with content. This approach is increasingly the trend for both consumer and enterprise applications.[2] The relationships between people and other objects, whether these other objects represent other people or items of content, can be encoded, as depicted in Figure 3-4, in the form of actor-edge-object "triples," whereby the edge denotes an action or a type of

---

[2]The Microsoft Graph is a current notable enterprise example: https://developer.microsoft.com/en-us/graph/docs/concepts/overview

relationship and the object can represent another person or an item of content. So, for example, a user, Jane, viewing a document, could be represented as the behavioral-based triple, Jane-Viewed-Document1. Or, as another example, Jane might follow Tom to receive Tom's social-networking posts, which could be encoded as Jane-Follows-Tom (Tom is the "object" for this particular actor-edge-object triple, but could also be the "actor" in other triples). Collaborations within an organization can result in large numbers of these types of actor-edge-object triples (along with auxiliary information associated with the triples, such as timestamps), and such triples constitute the informational raw material that fuels anticipatory computing.

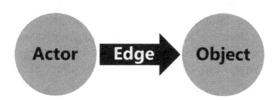

Figure 3-4. The actor-edge-object triple

The two examples of actor-edge-object triples just described are ones in which the edge represents actions or relationships that are explicitly performed by a person; i.e., viewing and following. But edges can also encode *inferences* about relationships between people and other people or objects. And while the performing of inferences is properly an activity of the subsequent Search and Discovery or Predictive Analytics elements of the data-to-learning-to-action chain rather than the Information Management element, I will discuss some important types of inferences here because the inferential results are so tightly bound to the information-management structures.

For example, an *interest graph* for each user can be generated. An interest graph comprises a set of interest-based relationships between a person and other things; for example, a topical area. These interests can be explicitly indicated by the user, but more often are inferred from other actor-edge-object triples that encode various user actions, such as content publishing and modifications, views, likes, following, geographic location information, and so on. Inferential algorithms place differing levels of weight on these behaviors and adjust the edges of the graph accordingly. For example, a user viewing a document that is associated with a certain subject area by itself is a weak signal of interest in the subject. However, contributing content that is associated with the subject is a relatively stronger signal of interest because of the additional user effort required. On the other hand, if the user makes hundreds of views of documents that are related to the subject, the volume of these signals can make up for the relative inferential weakness of each of the individual behaviors.

So, in general, an interest graph is constructed and adjusted based on the relative inferential value of each type of behavior and the volume of each type of behavior. Whether explicitly indicated or implicitly inferred, interests are generally a matter of degree, and so the interest relationships are represented as a continuum rather than a binary, yes or no, relationship.

The interest graph, often in combination with the social graph, enables an array of anticipatory functionality. For example, an item of content can be suggested based on the specific degree of interest a user has with respect to topical areas that are associated with the item of content. Or another person can be suggested to a user based not only on the user's social graph, but also by accounting for an inferred similarity of interests between the user and the other person. Most fundamentally, an interest graph provides the capacity for an automatic, fine-grained personalization that automatically adapts as a user's interests and preferences change over time. High-quality personalization enhances learning, which can, of course, positively affect decision making.

Another relationship between people and topical areas that is particularly relevant for organizational applications is expertise. As in the case of interests, the expertise level of a person with respect to a subject can be explicitly indicated by the person (e.g., within structured profiles) or by other people (e.g., endorsements), or it can be inferred from the person's actions and/ or from other people's interactions with the person's content contributions. While inferences of expertise can be derived from much of the same behavioral information as inferring interests, the inferential methods differ. For example, a user's view of a document related to a subject indicates a bit about the user's interests but nothing about their expertise (unless we happen to know something about the expertise level of the document, which is sometimes possible, and potentially by inferential means). But if a user publishes content related to that same subject that is positively interacted with (e.g., liked or rated highly) by other users, that may well tell us something about the person's expertise with respect to the subject, and more weight may be given to positive interactions by those users who are already inferred to have relatively high expertise levels.

Just as the social graph is continuously updated as people add or modify connections, the interest and expertise graphs are continuously updated based on inferences from people's actions. Taken together, these social, interest, and expertise graphs lie at the very heart of the anticipatory systems that we can expect to be working with going forward, whether as consumers or within our organizations—systems that automatically and continuously learn from and adapt to us, thereby accelerating *our* learning by anticipating and delivering to us the knowledge and expertise we need right when we need it.

In fact, anticipatory computing is an example of a powerful capability that is made possible by innovations spanning nearly the entirety of a data-to-learning-to-action chain. First, it rests on the acquisition and filtering of a source of data that was previously generally ignored, particularly in enterprise settings: incidental user behaviors. Second, the architecture of learning is the approach to managing information that has emerged to facilitate anticipatory computing. Third, the predictive analytics that generate inferences about user interests and expertise is also a recent innovation. And finally, at the interface with the user, search and discovery becomes a more personalized experience, and increasingly in a more conversational form.

Neural Networks

Neural network–based technology, particularly neural networks with many layers (i.e., *deep learning*), have made remarkable strides in the past few years in a variety of application areas, most notably in image recognition, but also in areas as diverse as language translation and games of skill, such as poker and Go. Their versatility and rapid advances suggest that they will no doubt play an increasingly important role in a wide range of enterprise applications.

The graph-based structures we have so far discussed in this chapter should not be confused with neural networks, however, which, while they have similar general structures, have very different modes of operations and, typically, application areas. Neural networks also require special types of information structures that serve as training inputs and from which the neural networks automatically learn, via either a supervised or an unsupervised mode. These neural network *training structures* are appropriately considered a part of the Information Management element of the data-to-learning-to-action process, while the actual *operation* of the neural network more closely corresponds to the Predictive Analytics element. For example, for neural networks that learn to identify objects in images, training sets of labeled images are required—a specialized structure within the Information Management element.

It is noteworthy that with the advent of neural network training sets we are witnessing a transition in the Information Management element of data-to-learning-to-action processes from information that is solely structured to facilitate human learning to that which also facilitates machine learning. While these structures are *currently* generally different in nature, it seems inevitable that that as neural network technology continues to evolve, human-oriented information structures will also be sufficient for neural network–based learning.

Information, Knowledge, and Learning

Before leaving the Information Management element, we should address the question of the distinction between information and knowledge that often arises. Exactly what is meant by knowledge as distinguished from information in common parlance, as well as in the more formal literature, is somewhat murky, and so I prefer to simply depict information and knowledge management combining to overlap multiple elements of the data-to-learning-to-action process, as was depicted in Figure 3-2. This aligns with the reality of the broad range of features that are often included in systems that would commonly be considered knowledge-management systems—examples of which we will discuss in more detail in the next chapter when we map various IT applications to the data-to-learning-to-action chain.

Of course, there is a natural sense that knowledge is a more organized form of information, perhaps at a higher level of abstraction, and is closer to being in a usable, actionable form for supporting decision making. Knowledge that is of particularly high value is commonly termed *insights* or *wisdom*. Notice that while these terms can be hard to crisply distinguish, what *is* apparent is that all the terms in the chain of transformations from data to information to knowledge to insights and wisdom are *nouns*. *Learn* is the *verb* that does the transforming of the nouns into one another along that chain.

Data, information, knowledge, insights, and wisdom are *nouns*. *Learn* is the *verb* that does the transforming of the nouns along the data-to-learning-to-action chain.

Search and Discovery

The Search and Discovery element of the data-to-learning-to-action process pertains to the access of the managed information of the prior Information Management element, either by an intentional method, typically performed by a human user (search), or by an automatic surfacing of information (discovery). Note that the term *discovery* here, which connotes a computer-generated suggestion or recommendation, should not be confused with the legal process of discovery, a function that is more aligned with the Information Management element of the data-to-learning-to-action process.

Search is distinguished from discovery in that a search function takes an explicit input, such as, for example, a word, phrase, or image, to guide its surfacing of information, whereas a discovery function does not. Search can therefore be considered just a special type of discovery system in which a more explicit signal of intentionality is provided to the discovery system. In both cases, fundamentally, the system makes *inferences* derived from the information that forms a basis for the information that the system in turn delivers. The information that is delivered by search and discovery systems may include documents, music, images, applications, and other people. And whether for search or discovery, the inferences that are applied may be those that are embedded within an architecture of learning-type structure, as was discussed in the Information Management section of this chapter. The inferences may be further enhanced by leveraging knowledge bases of "common sense" knowledge, i.e., facts about the world, or what is, again, commonly referred to as *semantic understanding*; for example, understanding relationships such as baseball is a sport, players swing a bat in baseball, the bat is often made of wood, wood comes from a tree, and so forth. With enough facts and the ability to chain facts together, sophisticated deductions can be made by a system, which enables more subtle, indirect types of inferences than is possible from user behavioral information alone.

A recommendation engine that infers the preferences or interests of users and delivers recommendations based on the inferences is an example of a discovery system. The recommender system will typically base its recommendations on inferences that are derived from behavioral information, such as historical patterns of interactions with content by the recommendation recipient and/or other users. The recommendation system may further make its recommendations based on the *context* of what a user is currently doing—for example, what a user is currently looking at online, where the user is currently physically located, what or who the user is currently physically proximate to, and so on.

Search and discovery systems apply a variety of types of algorithms to perform their inferences. These may include elaborate scoring systems in which, for example, documents are scored based upon the degree to which the document's text matches an entered search term, based upon user behaviors directly associated with the user or people inferred by the system to have similar characteristics of the user (i.e., "collaborative filtering" techniques), based upon the respective inferred expertise levels of the document and recipient of the suggested document, based upon whether the user is inferred to already be familiar with the contents of the document, and so on. These inferences may be performed on the fly by the search and discovery systems and/or they may leverage inferences that are embedded within a graph structure and may be supplemented with semantic-based capabilities. Increasingly, modern neural network–based techniques, such as deep learning, are being applied, both for the matching of search terms or queries to candidate content as well as for making inferences from behavioral information.

Some level of personalization, whereby the inferences are tuned to the specific context and behavioral history of a user, is now the standard for both search and discovery. As already mentioned, personalized search and discovery can be considered part of the overall field of anticipatory computing, whereby the preferences, interests, and even current expertise levels of a user are inferred and anticipated by the system on a continuous basis. In general, modern search and discovery is designed to surface what is relevant to users while suppressing that which is not relevant, thereby addressing the ever-increasing problem of information overload. In other words, the goal is to deliver signal from an inherently noisy environment, resulting in more efficient learning.

There is a trade-off however: personalization, while certainly enhancing human learning and productivity, when taken to an extreme can promote a counter-productive "filter bubble" problem.[3] A filter bubble connotes a self-reinforcing personalization that promotes an ever-increasingly deeper, but also narrower, perspective. That can have unfortunate consequences for innovation, which typically requires the *combining* of concepts from *diverse* areas and perspectives. So, an important direction for search and discovery functionality is to get the trade-off right: retaining useful personalization, but also having a capability for delivering beneficial serendipity by identifying information and perspectives outside of the user's typical interest patterns and incorporating them in search results and recommendations.

The user interfaces for search and discovery applications are also becoming more sophisticated. Natural language–based digital assistants such as chat-bots are rapidly supplanting the good old-fashioned typing of a key word by a user, and conversation-based system interfaces are clearly destined to become standard. Combining natural-language conversational capabilities with common-sense understanding about the world along with personalization based on user behavioral history is an inevitable trend. Search and discovery capabilities seem destined to be primarily embodied within such intelligently adaptive conversational agents.

Increasingly it is desired that search and discovery applications also provide an explanation for *why* the system delivered the information it did, which can provide the recipient of the information additional insights and can promote more trust in what would otherwise be a rather opaque process. In fact, the desire to overcome the opaqueness of inferential-based search and discovery systems has led to the right for a user to receive an explanation for the system's decision to be a legal requirement in some jurisdictions.[4] Delivering succinct but meaningful explanations is a challenge given the complexity of algorithms that are applied in search and discovery systems, and the more wide-spread application of deep learning makes it that much more difficult. Nevertheless, explanatory capabilities are highly desirable, and may even be required for some applications, so they will necessarily be an important aspect of search and discovery going forward.

---

[3]Pariser, Eli. *The Filter Bubble: What the Internet Is Hiding from You*, Penguin Press (New York, May 2011).
[4]http://www.slate.com/articles/technology/future_tense/2017/05/why_artificial_intelligences_should_have_to_explain_their_actions.html

Predictive Analytics

Predictive analytics is the essence of learning. Learning, whether by mind or machine, is most fundamentally a process of reducing uncertainty, or, in other words, increasing predictive accuracy. This predictive accuracy may be thought of in quantitative terms, such as probabilistically, or in more everyday terms, such as having greater *insight* into what will happen. Even more directly than in the case of Search and Discovery, the Predictive Analytics element is about determining signal that lies within the data and information noise. And whether by mind or machine, for decisions in the face of uncertainty, a prediction, whether implicit or explicit, must necessarily be made prior to performing the action.

This is the stage of the data-to-learning-to-action process on which data scientists are particularly focused by making meaning of data, inferring causal relationships, and predicting future events using a variety of analytical tools and techniques. Statistical and probabilistic techniques are typical approaches. Data clustering is often an important technique for identifying patterns and generating insights. This is also the stage in which artificial neural networks are trained and then applied to make identifications and to reveal patterns.

As we discussed earlier, it is also the element in which structures associated with the Information Management element, such as enterprise graphs, may be updated. For example, nodes may be added or deleted, and relationship weightings may be adjusted based upon predictive inferences. The predictive inferences may be derived from analyzing user behavioral information, but may be also be supplemented by applying semantic-based capabilities to better "understand" the content.

Whether in computing-based structures or structures within our brains, this inferential learning is a continuing process because the inferences can keep getting better and better (i.e., uncertainty is reduced and predictive accuracy is increased) through the cumulative processing of a continuing stream of additional inputs. Figure 3-5 conceptually sketches this learning in action, whereby the existing inferences that are represented as weighted edges among the nodes are adjusted by the learning process. This contrasts with knowledge, which both serves as the *basis* and is a *result* of a learning process.

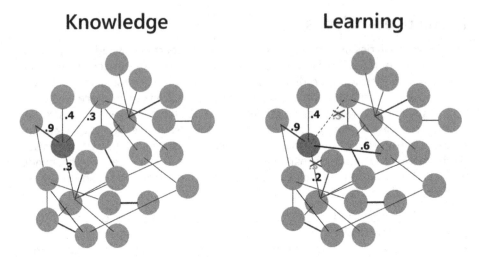

Figure 3-5. Learning in action

To put a finer point on this distinction, data, information, and knowledge are all *stocks*—that is, in theory at least, they can be measured at a point in time. Again, learning is different; it's a verb, not a noun, and can be considered a *flow* representing a change in knowledge over time and, more specifically, a change in probability distributions that are influenced by the additional knowledge. As I analogized in *The Learning Layer*, in financial accounting terms, knowledge is like the balance sheet at a point in time and learning is like the profit and loss statement for a given period. The results of the profit and loss for the period determine the balance sheet for the beginning of the next period. Or, in terms of a calculus analogy, integrating over a learning function results in knowledge.

In many cases, a learning system, as represented conceptually in Figure 3-5, is also *recursive*, in that learning that occurs over a period yields a subsequent knowledge base that in turn influences subsequent learning. As an example of this, a graph structure that is altered by a learning process can influence subsequent user behaviors that are in turn used as a basis for new inferences that again alter the structure, and so on.

And while most of this brief discussion of the Predictive Analytics element has focused on algorithmic and machine-based predictive capabilities and approaches, we should bear in mind that the reality is that much of the predictive analytics that are applied in everyday business situations and will continue to be applied are based on that wonderful non-silicon neural network technology, the human brain!

Process and Collaborate

Process and Collaborate is the stage of the data-to-learning-to-action process in which people make additional meaning—or, in other words, learn from— the information and knowledge provided by the previous stages. Process and Collaborate is an element that is more oriented toward human decision making; for automated data-to-learning-to-action chains, the learning flow may go directly from Predictive Analytics to Decide and Act.

The *process* part of the element includes the developing of models and/ or packaging information for colleagues' consumption to facilitate decision making. So, for example, spreadsheet-based modeling, preparing documents and presentation materials, and preparing and operating dashboards are examples of activities of the process aspect of this element.

The *collaborate* aspect of this element relates to collective interactions and learning among individuals who share a common purpose, laying the groundwork for making group decisions. The collaboration may be performed through functions such as email, social networks, audio and/or video conferencing, online chat, or face-to-face meetings and live events (Post-it notes can still be invaluable!). The collaborative activities may be organized in the form of an explicit project-management process. Going forward, AI-based conversational agents will undoubtedly become important at this stage of the data-to-learning-to-action process, as Process and Collaborate will no longer be the exclusive province of people.

However, regardless of the technology applied, the Process and Collaborate element is where human judgement most comes into play, including the special dynamics of group deliberations and decision making. While the human brain is a marvelous decision-making system, it certainly has well-documented cognitive biases and deficiencies.[5] The danger is that even if wonderfully sophisticated, leading-practice approaches are applied in the upstream elements of the data-to-learning-to-action process, decisions can be botched by the vagaries of human decision-making at this late stage. The good news: that which is adequately understood can be improved![6]

---

[5]https://betterhumans.coach.me/cognitive-bias-cheat-sheet-55a472476b18
[6]Bang, Dan, and Chris Frith, "Making better decisions in groups", *Royal Society Open Science*, August 2017. http://rsos.royalsocietypublishing.org/content/4/8/170193

Decide and Act

We are finally at the stage of the data-to-learning-to-action process at which decisions on potential actions are made! These may be individual decisions or collective decisions. We usually think of decisions being made by people, and, of course, that is true, but decisions may also be made directly by systems, and increasingly so as technology progresses.

Decisions are made with respect to potential actions, but what actions? It is important to understand the full range of actions that may be decided on to ensure that the most potentially valuable actions are considered, and that may well be a different set of actions than just the actions that have been previously considered for a decision. So, an element of creativity should be applied in thinking about alternative actions. Furthermore, we say decisions are made with respect to potential actions, but we need to be careful to take the term *decision* to be robust enough to include the choice of not taking any action or to defer taking an action.

Actions may include acts by an actor that directly affect the external world as well as those that are directed back to the actor itself. Some actions are explicitly intended to generate useful data, which, as mentioned earlier, are actions we generally call *experiments*. As with the other elements of the data-to-learning-to-action process, actions may be performed by people or machines.

Considering alternative actions is an important opportunity for value creation and benefits from innovative thinking.

As depicted in Figure 3-6, actions, whether explicitly experimental or not, may result in new data that can then be acquired and further processed by the other elements of the data-to-learning-to-action process. In such cases, data-to-learning-to-action chains can be considered a part of a recursive, closed-loop process—a feedback loop that results in a powerful self-reinforcing learning effect.

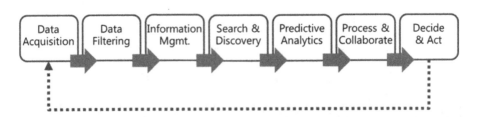

Figure 3-6. The data-to-learning-to-action loop

Actions have a net value, which can be negative as well as positive. Part or all the net value of an action may be attributable to *learning value* (which is necessarily always positive), and, as we will see in subsequent chapters, we can quantify the value of an action in financial terms, i.e., monetary value. An action has learning value if the action results in new learning that can potentially affect a subsequent decision. And as I have stressed, this same universal learning process, whether recursive or not, is applicable to both people and systems, as well as mixes of people and systems operating together across the elements of the data-to-learning-to-action process.

Data-to-learning-to-action chains can also be nested. That is, there can be subsidiary data-to-learning-to-action chains within one or more elements of another data-to-learning-to-action chain. Take, for example, decisions on product marketing. Within the Data Acquisition element of the overall data-to-learning-to-action process associated with the product marketing decision, there may well exist a subsidiary decision on what sources of data on consumer preferences should be acquired, along with an associated data-to-learning-to-action process for that subsidiary decision. The results of these nested data-to-learning-to-action chains, some of which may be self-reinforcing loops, as well as combinations of minds and machines, is what leads to the complexity, but also the vibrancy and dynamism, that characterizes modern organizations!

Summary

In this chapter, we briefly reviewed each of the elements of the data-to-learning-to-action chain for organizations: Data Acquisition, Data Filtering, Information Management, Search and Discovery, Predictive Analytics, Process and Collaborate, and Decide and Act. The overview touched on current states of the elements, as well as trends, with an overriding trend for all the elements being an accelerating transition to more adaptively intelligent systems. We noted that learning traditionally would be considered to correspond more closely to the elements of the chain that are farther downstream, but that the capability for learning is increasingly creeping backward along the chain; for example, being directly embedded in information structures. This trend reinforces further the data-to-learning-to-action perspective that learning should be considered a flow along the entire chain.

Tech Stuff and Where It Fits

I have emphasized and will continue to emphasize that optimizing data-to-learning-to-action is certainly not just about technology—it's about people, process, and technology. Nevertheless, technology is a massive investment area for any organization. Its rapid evolution is so highly dynamic that technology necessarily presents continuing decisions for any organization, and it can disrupt entire business models in addition to individual processes. It therefore necessarily demands thorough and continuing attention.

Further, the sad reality of enterprise IT (and when I refer to *enterprise* IT or applications, it is meant to equally apply to other types of organizations, such as non-profits and governmental) is that it is seemingly particularly prone to shaky decision making. Perhaps that is at least partly due to the complexity and confusion factors discussed in the first chapter that can so befuddle. In some cases, it may be the result of the organizational structure, or the way decisions in general are made regarding IT, or because of an underestimate of the change-management aspect of implementing new technology. Whatever the factors, the result has been that it is not unusual for enterprise applications to be purchased and relegated to a fraction of the expected deployment target areas of the organization, or to be completely shelved without any deployment at all. On the other hand, it has also not been unusual for IT that would have otherwise delivered significant value to not be acquired at all, or to be acquired later than when it would have had the greatest positive impact. The fact that the traditional ways of making decisions on IT have so often

© Steven Flinn 2018
S. Flinn, *Optimizing Data-to-Learning-to-Action*,
https://doi.org/10.1007/978-1-4842-3531-7_4

resulted in these value-destroying outcomes reinforces that it necessitates our attention here (and remember, software is accounted for as an asset; if it isn't adding sufficient value, it contributes to that underperforming ROA problem).

So, in this brief chapter, we will examine how selected categories of information technology map to the data-to-learning-to-action process. This serves two purposes. First, it will provide some additional perspectives on the elements of the data-to-learning-to-action chain beyond our tour of the previous chapter. And second, it will provide some quick coverage of various IT application areas that will be relevant when we delve into patterns of constraints on data-to-learning-to-action value and their potential solutions in subsequent chapters.

Of course, we can only touch on a fraction of the popular technologies and their associated mappings to the various elements of data-to-learning-to-action processes. Some that we will touch on are general purpose in nature, while others are more specific to functional areas such as human resources and R&D. The focus will also be more on IT that is positioned at the higher levels of abstraction in the "IT stack"—typically applications that people directly interface with and that directly support their decision making—rather than lower-level structures and technologies, such as the underlying information structures that we discussed in the last chapter, or communications infrastructure, general cloud-computing infrastructure, and the like.

And there are, of course, vast numbers of more narrowly focused and often more cutting-edge application areas that we cannot cover during this brief overview. These have been particularly facilitated by cloud-based computing, as well as by application programming interfaces (APIs) that enable integration with large, established applications, as well as with other more narrowly focused applications.

It is important to understand, and this will be highlighted during our tour, that many common technologies, particularly enterprise software applications, naturally evolve to become quite sprawling in nature, spanning multiple elements of data-to-learning-to-action processes. But what is ultimately important for optimizing the data-to-learning-to-action process are specific *features* that map to specific elements of the process. The constraints on value in a data-to-learning-to-action process necessarily need to first be examined at that feature-based level of granularity so that value can be meaningfully quantified. Only then can specific technology options, most often comprising bundles of features, be properly evaluated as potential means to resolve value bottlenecks.

In the following sketches, only the *core* features of the selected technologies are highlighted against the data-to-learning-to-action chain; in practice, marketplace solutions in these technology areas may include features that span other elements, at least to some degree, as well. This reflects the sprawl factor just mentioned, a consequence of established applications' seeking to expand their scope of functionality in accordance with a strategy of capturing an ever-greater share of technology budgets. Another dynamic that is clearly at play is the extraordinary ascent of intelligent, machine learning–based capabilities, making it an increasingly mandatory feature enhancement for many application areas. The result is that the Predictive Analytics element of the data-to-learning-to-action process is increasingly embodied, to at least some extent, in application areas that just a few years ago were devoid of any such capabilities.

We'll start our tour with several general-purpose applications that apply to just about every organization and then move on to some of the major applications that are more specific to various functional areas.

General-Purpose Applications

Of the following applications that we will cover, document and content management and business intelligence are currently found in nearly every organization. The third cross-functional application that we will discuss, enterprise social networks, is not yet as common but is rapidly becoming a standard part of enterprise collaboration.

Document and Content Management

Document-management systems enable the publishing and retrieval of documents and typically come equipped with permission settings, version controls, and auditing capabilities. Increasingly, document management may comprise a set of features that reside within broader-featured products that include other capabilities such as collaboration and process management. The core features of document management are centered on the Information Management element of the data-to-learning-to-action process and usually have built-in search capabilities, and have increasingly come with automated suggestions so that there is some overlap with the Search and Discovery element as well, as shown in Figure 4-1.

Figure 4-1. Document- and content-management systems

More broadly, *content-management* systems provide similar functionality as document-management systems, but with expanded capabilities to include other types of content in addition to documents, including images, video, audio, and even software applications (apps). Document- and content-management systems form the heart of what are commonly referred to as *knowledge-management systems*.

Business Intelligence

Business intelligence systems organize and package information to assist people in making decisions. These systems often include functions that facilitate data mining, creating models and performing analyses, and displaying the resulting information in report and/or graphical formats for consumption by decision makers. In the past, the humble electronic spreadsheet bore much of the burden of the business intelligence needs for most organizations. Now, more sophisticated systems commonly include advanced data-query methods, statistical modeling, and graphical display features. These technologies therefore most closely align with the Search and Discovery, Predictive Analytics, and Process and Collaborate elements of the data-to-learning-to-action process, as shown in Figure 4-2.

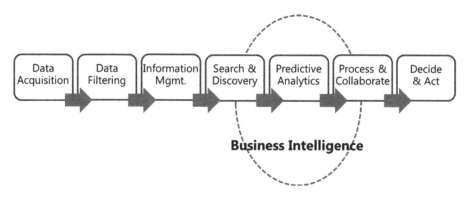

Figure 4-2. Business intelligence systems

Increasingly, the line between business intelligence and other enterprise application packages is blurring as those packages build in more predictive analytic and general decision-support capabilities.

Enterprise Social Networks

Enterprise social networking is a key collaboration technology that has been transferred from the consumer environment (e.g., Facebook) to the enterprise, supporting cooperative enterprise communities. Its advantage versus, for example, email is that it is network-based, is people-centric, and enables a pull rather than push model—that is, users can subscribe to content associated with individual users and topical areas. These features enhance the ability for colleagues to effectively and efficiently collaborate on projects and individual items of content, as well as enhance the discovery of people with expertise that fits with expertise needs.

Enterprise social networking maps to the Process and Collaborate element of the data-to-learning-to-action process as depicted in Figure 4-3. Its fundamental feature of subscribing to activity streams that are of interest is increasingly bundled with other collaborative functionality such as real-time messaging and video conferencing, thereby delivering enhanced enterprise collaboration functionality, as well as becoming a capability that is increasingly merged or tightly coupled with document- and content-management functions.

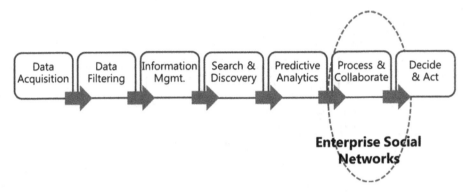

Figure 4-3. Enterprise social network systems

Function-Specific Applications

In this section, we will briefly examine some of the larger applications that apply to functional areas such as R&D, sales and marketing, supply and manufacturing, and human resources.

You may have noticed that the general-purpose applications discussed earlier tended be more oriented to the downstream elements of the data-to-learning-to-action chain. That's because functionality that is relatively closer to the decision making, such as knowledge management, analytic capabilities, and collaborative platforms, can generally be effectively leveraged cross-functionally. On the other hand, data tends to be specific to functional areas, so the following function-specific applications are more apt to map to the upstream elements.

But because of the ascendancy of intelligent analytics, particularly machine learning–based analytics, coupled with the economic pressure to increase the scope of commercial packages, many of the major function-specific applications map to significant portions of *both* the upstream and the downstream elements of the data-to-learning-to-action chain.

R&D: High-Throughput Experimentation Technologies

In the fields of pharmaceuticals and materials science, automated methods of performing experiments that can perform orders of magnitude more experiments per unit of time than can manual methods are increasingly being applied to accelerate learning. These are typically robotic-based technologies, which generate data that can then be processed by subsequent stages of the data-to-learning-to-action process (see Figure 4-4). It is often the case that any given individual data point generated in this automated fashion is less valuable than data points resulting from old-fashioned, manually configured experiments because of the "shotgun" effect of the automated method. But the sheer volume can make up for the high proportion of "throw away" results, thereby accelerating the learning process that results in making useful discoveries. This "fail early and often" approach flips the old admonishment of "haste makes waste" on its head. It is now actually the case—with the very low cost per unit of experimentation and the accelerated time-to-result that these automated methods can provide—that today's realistic perspective is "waste makes haste!"

Figure 4-4. High-throughput experimentation systems

This same low-cost-of-experiments phenomenon that is transforming physical product development is also at work in other environments and represents a general economic trend. For example, for online systems, the A/B testing of functionality and interface options, which is revolutionizing the way computing interface decisions are made, can be considered a high-throughput experimentation technique.

And even more broadly, in our next sketch we address marketing analytics, an area that is being transformed by the ability to easily and at low cost attain massive amounts of consumer data, and often in real-time, derived from consumers' interactions with online assets.

Sales and Marketing: Marketing Analytics

Marketing analytics systems collect data from marketing campaigns that target channels such as email, social networks, and mobile devices and then enable the derivation of predictive insights from the data that support marketing- and strategy-related decisions. This is an application area that was once mostly about data gathering and then relying on the labor-intensive making of meaning of the data and/or leaving the analysis of the data to other tools. It is now following a pattern that is common to other application areas: moving toward highly integrated systems that feature built-in machine learning and sophisticated statistical capabilities, resulting in contemporary marketing analytics systems that have a broader expanse across the data-to-learning-to-action process, as is illustrated in Figure 4-5.

Figure 4-5. Marketing analytics systems

Sales and Marketing: Customer Relationship Management

Customer relationship management (CRM) systems help organizations make decisions about how to best market and sell to potential and existing customers, as well as decisions related to customer retention. In general, CRM functions can be broadly characterized as either marketing-related or sales-related. The typical capabilities of customer relationship management systems include the ability to capture and manage information about customers and prospects, tracking steps in marketing and sales processes with respect to customers and prospective customers, and reporting and analytic functions that support decision making. In addition to tracking financial-related expectations and outcomes with respect to customers, metrics related to how satisfied a customer is with their relationship to the supplier are tracked. Figure 4-6 illustrates a rough mapping of CRM systems to the data-to-learning-to-action chain.

Figure 4-6. Customer relationship management systems

In practice, there is significant overlap of CRM systems with marketing analytics and business intelligence applications that have already been described. As with other major enterprise application areas, this overlap tends to continuously increase as vendors attempt to expand the feature scope of their offerings.

Supply and Manufacturing: Supply Chain Management

Supply chain management systems cover a broad spectrum of functionality that can include, for example, vendor management, procurement, logistics, inventory management, production-workflow management, and demand management. These capabilities support decisions such as what products and services should be procured from what vendors, how to best transport these products and services to where they are needed, how much of the products and services should be acquired, as well as determining the inventory levels of materials that should be maintained. Most fundamentally, supply chain management seeks to facilitate optimal operational decisions given the uncertainties that are associated with both the supply and the demand for products and/or services. The ability to reduce those uncertainties constitutes high-leverage learning opportunities since the benefits ripple through multiple aspects of the business.

While always an area in which modeling and optimization methods played a significant role, as in other application areas, supply chain management systems have been evolving to become even more predictive analytics–intense, with increasingly sophisticated models. This is in addition to handling the more basic functions of managing significant amounts of supply chain–related data and information, as is illustrated by Figure 4-7.

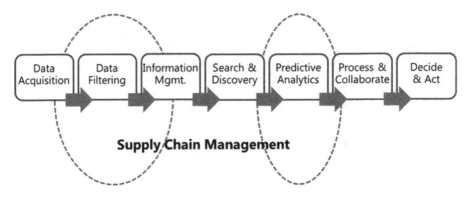

Figure 4-7. Supply chain management systems

Human Resources: Talent Management

Talent management systems help manage an organization's human resources. They include capabilities for recruiting, or more generally, talent acquisition, skill and expertise development, and performance management and support a variety of decisions related to the acquisition, development, and retention of human capital. The skill- and expertise-development capabilities may include those traditionally found in stand-alone learning management systems. As is the case for many of the other systems sketched out in this chapter, enhanced analytical capabilities, including machine learning–based features, are inevitably and continuously being included in talent management systems.

As just one example, expertise levels are beginning to be measured based on what people do, not just what they say about themselves. As we discussed in the last chapter, inferences of expertise levels of a person can be derived based upon the levels of expertise of the people who positively interact with content that is authored by the person.

Figure 4-8 depicts a rough mapping of talent management systems to the data-to-learning-to-action process. Recruiting aspects such as candidate interviews and résumé searching map more closely to data acquisition and filtering elements, while skill and expertise management more closely map to the information management and analytics-based elements of the data-to-learning-to-action process. Of course, performance management inherently includes some degree of collaboration, which is not explicitly shown on the sketch. In practice, cross-functional systems that are more specifically designed for collaborative interactions are more likely to serve as the platform for detailed, ongoing performance feedback.

Figure 4-8. Talent management systems

As mentioned, this brief tour of enterprise IT touched on only the major application areas that most commonly apply to organizations. There are myriad other applications that are more detailed "point solutions," or applications that tend to fall in the white space between the larger application areas; for example, ideation systems that help foster creativity and innovation in an organization by systematizing the process and fostering greater grassroots participation, as well as encouraging beneficial serendipitous interactions.

Or, for example, human resource analytics, which is an emerging analog to marketing analytics but with a focus on an organization's employees rather than its consumers, enabling a better understanding of the people who work in the organization and facilitating better decisions with respect to them. An example of such analytics that we have already discussed is expertise inferences. Inferring an employee's level of *influence* on other employees (i.e., *applied* expertise) is another example of employee analytics, as is employee *sentiment analysis*.

But regardless of the application, whether broad or narrow in functionality, the *features* can be mapped to the data-to-learning-to-action chain! And it is the feature level that is relevant for addressing learning constraints and for which the resulting expected value of implementing the feature can be determined, as we will outline in subsequent chapters. This feature-specific value can then be aggregated as appropriate to determine the expected net value of implementing an overall application package that embodies the features.

Summary

The following are some take-aways from our brief tour of common enterprise technologies and their rough mappings to our data-to-learning-to-action process:

- People, process, and technology, in that order, are key to business-performance improvement, but the extraordinarily rapid advances in enterprise applications and associated required investment levels necessitate our significant focus on the technology.

- The reality is that investments in enterprise information technology have been fraught with less than optimal decision making, with the result being either the acquisition of systems that are never used or are under-utilized, or the failure to acquire systems that could make a significant difference but that are not acquired and implemented in a timely manner. This reality is another reason IT must necessarily be a significant focus of just about any data-to-learning-to-action optimization.

- There is an increasing overlap in the functionality of major application categories as application suppliers seek to expand the scope of their products. At the same time, however, although we could not cover them in this brief review, there is a continuous emergence of more focused, leading-practice solutions, often enabled by integrating with other recently emerged applications, as well as with larger, well-established applications.

- Intelligent analytics, including machine learning–based functions, are rapidly being integrated with nearly every major enterprise application area.

- Our mapping exercise reinforces that technology decisions must be made from the finer-grained element-level perspective of the data-to-learning-to-action process and in the context of people, process, as well as technology, rather than at the level of commercial products. A decision at the commercial-product level must necessarily be informed by the finer-grained perspective.

Now on to the methods for determining, among other things, if these and other technologies are part of the problem or part of the solution in optimizing a data-to-learning-to-action process!

Reversing the Flow: Decision-to-Data

As we have seen, learning can be thought of as flowing along the data-to-learning-to-action process in the sense that uncertainties that are ultimately embodied, implicitly or explicitly, as probabilities are continuously adjusted as the process moves forward. It is clear, then, why learning is the key to business performance, since learning is the *process for becoming increasingly effective at predicting by reducing uncertainty*, and everything that *really matters* from a business-performance standpoint revolves around better actionable predicting. Decisions in which there are *no uncertainties* are amenable to rote rules. For example: If the widget is of type X, then perform action Y. There may be many of these types of decisions that occur all the time in an organization, but most of them simply become embedded in automated processes that by themselves have little *sustainable* impact on comparative business performance. On the other hand, a decision on *when* and *how* to automate those types of decisions *can* impact comparative business performance, and, of course, those decisions do require learning and the application of the universal process, data-to-learning-to-action, because there are uncertainties that affect those types of decisions.

© Steven Flinn 2018
S. Flinn, *Optimizing Data-to-Learning-to-Action*,
https://doi.org/10.1007/978-1-4842-3531-7_5

We also know that since learning can be thought of as a flow, the application of our theory of constraints-based thinking implies that there will inevitably be bottlenecks that constrain the learning flow. More importantly, there will be constraints on the *value* of the learning throughput of the data-to-learning-to-action process. So, alleviating the constraints on the throughput of data-to-learning-to-action processes is clearly the prescription for improving business performance.

But the reality is that there will be many data-to-learning-to-action processes active for any organization. There will be some that are specific to functional areas, such as manufacturing, finance, HR, and so on. There will even be data-to-learning-to-action processes nested inside other data-to-learning-to-action processes. Once we choose a data-to-learning-to-action process to optimize, we have a method to guide us, which we will begin to lay out in more detail later in this chapter. But how do we *choose* which data-to-learning-to-action processes to target to begin with?

In some cases, it may simply be obvious; for example, when there is by common consensus an important process or functional area that is clearly broken, and it is commonly accepted that something needs to be done about it, even though it may not be clear exactly how. In such cases, it *may* make sense to immediately begin examining the associated decisions that are being made, and then apply the methods that are presented later in this and ensuing chapters. But I say "may" make sense because a quantified value check of the type discussed next should be performed to scale appropriately the attention and effort that is directed to the problem.

More generally, for situations in which overall organizational performance needs to be improved and for which there are many possible targets for trying to achieve that improvement, it is imperative to apply a structured approach for prioritizing the targets. And for that, we need to understand the *value drivers* of the organization.

Value Drivers

Ultimately, what matters for prioritization is the expected impact on the organization's long-term financial performance.[1] After all, an important motivation of the optimizing data-to-learning-to-action approach is the underper-

---

[1] This is assuming a traditional for-profit organization. Other types of organizations may have fundamental objectives other than financial. And for-profit organizations may have other objectives in addition to financial objectives. The methods in this book can still apply for such non-financial objectives if there exist, or can be developed, quantified metrics that can be applied to measure the success in achieving the objectives. Those metrics simply replace monetary value as the proxy for the *utility* that is to be maximized by the associated learning and decisions.

formance of businesses as measured by metrics such as return on assets. And those are just aggregate performance metrics across companies—some organizations will be in even worse shape, and even those that are in better-than-average shape almost surely will have certain divisions or functional areas that are underperforming. So, it is incumbent on us to be guided by aspects of overall financial performance in prioritizing what data-to-learning-to-action processes we are going to address first. And applying *value drivers* is the way to do that.

What we mean by *value drivers* are the key activities that map directly to, and have the greatest leverage on, an organization's performance, typically as measured by its financial statements. The mapping to financial statements is generally straightforward, but determining the degree of leverage may require some sensitivity analysis.

Value drivers are those activities that have the greatest leverage on an organization's financial performance.

The financial statement that will generally be the fundamental starting point for value-driver analysis is the profit and loss (P&L) statement, which, at a high level, might look like some variation of Figure 5-1.

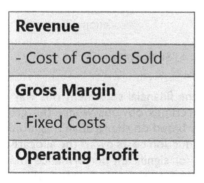

Figure 5-1. Basic P&L statement

What we want to do as a first step in identifying the drivers of value is to systematically decompose the key elements of the financial statement into the next level of detail, as shown in Figure 5-2. So, for revenue, we break out the next level of components, such as product volume and price per unit for a manufacturing company. For cost of goods sold, the next level of components for a manufacturing company might be various product-related cost elements, such as production costs and sales costs. Fixed-cost components would include direct fixed costs such as depreciation of the manufacturing facility in which the product is produced, as well as allocated fixed costs such as the overhead costs of various functional areas (HR, Legal, Finance, etc.) and management. For service-based companies, the same type of financial-statement decomposition holds, but, of course, the components will be different. In general, financial statements that are aligned with *activity-based costing* are most helpful to us during this exercise as they attempt to realistically link costs as much as possible to the revenue streams that the costs contribute toward, rather than applying more arbitrary allocation methods.

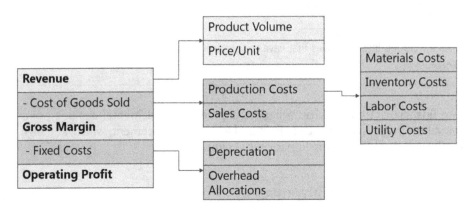

Figure 5-2. Decomposing the P&L statement

Simply breaking down the financial statement this way can be highly instructive—it often quickly becomes obvious where the greatest opportunity for improvement lies simply based on the relative magnitude of the various numbers. For example, if production costs dwarf the allocated costs of HR, which is the more obvious lever for significant performance improvement? Of course, production costs. This seems so simple, but too often executives simply orient performance-improvement attention and resources to what happens to be most expedient rather than to what truly matters!

This decomposition of the elements of the financial statements can continue to as fine a detail as is required, with the relative contribution to the bottom line (e.g., operating profit) examined for each branch of the financial-statement decomposition. For example, production costs might be further decomposed into costs of materials (such as ingredients), inventory costs, labor costs, and utility costs, as depicted by Figure 5-2. Costs of ingredients could be further decomposed into various types of materials used to make the products—such as for a confectionary product, the costs of the chocolate, nuts, sweetener, and so on. The decomposition is guided by the "follow the money" principle—if the impact on the financial statement is significant, the item should be broken down some more to determine the relative magnitude of the components.

For additional rigor, a sensitivity analysis is typically performed. This can be done by establishing ranges for the components and then determining the spread of the outcomes on the bottom line as we move from one end of the component's range to the other. So that the ranges represent a reasonable "art of the possible," they may be determined or calibrated by, for example, benchmarking results from other organizations in the same industry, or even from outside the industry. Evaluating the effects on, for instance, operating profit for these ranges can provide a reasonable understanding of a component's leverage on overall business performance.

The set of components that appear to have the most leverage on overall business performance are those that we term the value drivers. Our view of what value drivers we should be focusing our attention on may well shift as the analysis follows the various value branches from the base financial statements. For example, production costs might be identified as having significant value leverage, but further analysis might determine that materials costs are the dominant component of product costs, so it is the key driver of value. But it may be that the cost of chocolate is the dominant cost component of the materials costs, which is, in turn, the dominant component of production costs, so the cost of chocolate is really the value driver that we would want to focus on, as is illustrated by Figure 5-3. And as we will discuss, we will therefore want to particularly focus on the *decisions* that influence that ultimate value driver.

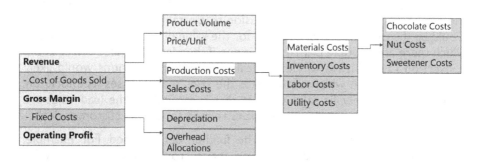

Figure 5-3. Determining the value driver

Other common examples of value drivers that we will touch upon in upcoming examples are product pricing and talent acquisition. These example value drivers represent fairly traditional financial-related items. But non-traditional aspects that affect business performance can also be value drivers and should therefore be carefully considered; for example, employee morale. This is an item that will not typically show up on traditional financial statements, but its effect on business performance may very well make it a value driver since employee *engagement level* has been identified as a critical factor for productivity.[2] Therefore, it should not be ignored by our process. What *is* required by our process is to be able to quantitatively translate the impact of the value driver, say, employee morale, to the business's finances. That will typically take some work and will likely include some subjectivity (and *subjectivity* should not be construed as a dirty word—it simply means applying *human-based* predictive modeling!), but it absolutely must be done, otherwise we risk working on the wrong things.

This process of determining value drivers and using them as a guide for where to focus business-performance improvement efforts is not new, although it is perhaps underutilized. What *is* new is that in our approach, rather than just applying various *ad hoc* methods to improve the drivers of value of an organization, we will systematically address improving those value drivers by getting right to the heart of performance improvement by identifying and improving the decision making—and the learning upon which the decisions are dependent—for the identified value drivers.

Decisions, Decisions

Associated with every value driver are one or more decisions that affect it. And echoing the process of decomposing financial statements into more detailed components in our quest to identify the value drivers, we must next identify the *primary* decisions that are associated with each of the value drivers. Then, these primary decisions may need to be broken down into their subordinate decisions, and so on. For example, if an identified value driver is the price for a product, then the decision that establishes that product price would be the obvious *primary decision* for the value driver. Or as an example in which there are multiple primary decisions, if the identified value driver is the conversion rate of leads to closed deals, the primary decisions would likely include decisions on which of the leads the sales team focuses its efforts on, decisions on the specific sales process that is applied for the leads that are selected, and negotiation-based decisions related to specific customer deals.

These primary decisions may then be further decomposed. For example, for the negotiation-based decisions on the customer deals, there may be sub-decisions on the deal's terms and conditions that are made by the legal organization, as well as for the pricing-related decisions that are made by the sales organization or management.

---

[2]Baldoni, John. "Employee Engagement Does More than Boost Productivity", July 04, 2013, https://hbr.org/2013/07/employee-engagement-does-more

To determine what our attention and resources should be focused on, we should always be guided by 1) the potential long-term *financial leverage* of improved decision making and 2) how *likely* it seems that we can *improve* the decision making. In our example, we may know that overall negotiation-based decisions are critical because they are tied to a value driver. But if the terms and conditions are rarely the show-stoppers in getting deals done, while pricing more often is, then pricing decisions clearly have more financial leverage, and our emphasis should therefore be on analyzing the pricing decisions and the associated data-to-learning-to-action process. On the other hand, if terms and conditions are often show stoppers and therefore have significant financial leverage, but the show-stopping occurs because certain terms are set by company policy and they cannot be changed within the scope of our improvement initiative, then there is no sense trying to address the associated decisions (although in such a case we should certainly try to pull the decisions on terms and conditions into the scope of the improvement initiative).

Working Backward

Now we are ready to work on improving a critical decision, one that is tied to a value driver (or that by consensus warrants immediate attention), and one that clearly has significant impact on the value driver. How do we proceed? We work systematically *backward* from the decision along our data-to-learning-to-action chain, as depicted in Figure 5-4. In a sense, we are reversing the flow of learning, learning *more* about the associated data-to-learning-to-action process and its learning flows by working in a step-wise manner back from the decision. We will do this in two passes. For the first pass, we will work back from the decision with the objective of just understanding the overall data-to-learning-to-action chain and the operations and learning flows associated with each of the chain's elements. We conduct this first pass so that we have the basic understanding needed to ask targeted, more detailed questions in a second pass that will home in on the limiting constraints on value in the chain.

Figure 5-4. Working backward along the data-to-learning-to-action chain

We also use this opportunity to build working relationships with participants who are involved with the various elements throughout the chain. This is key because much of what we will need from the second pass are thoughtful assessments by the participants, and we want them to be comfortable being forthcoming in their perspectives and assessments. There is usually little difficulty in establishing this trusting rapport because it is very often the case that no one has really solicited their candid opinions and assessments on the subjects of interest, and they therefore tend to very much appreciate the opportunity and are usually quite eager to relate perspectives that they may have had for some time but that never had a proper outlet.

Let's walk through some examples of this first pass. The element that lies immediately upstream of the decision is the Process and Collaborate element of the data-to-learning-to-action chain. So, we first analyze how the decision is made within the context of that element. For our pricing example, we might find that the pricing decision is made by the consensus of several people, including the sales lead for the customer account and several members of management, and that the decision is determined through communications conducted over email.

Having this basic understanding of the Process and Collaborate element in the context of the pricing decision, we then move backward to the next element of the chain, Predictive Analytics. Here, we examine the predictive assumptions that each of the contributors to the decision have. These predictive assumptions could be based on human intuition that comes with years of experience and/or they could be based on explicit models of, for example, customer price elasticity or competitor bid patterns, and we would want to attain a basic understanding of the predictive assumptions and implicit or explicit models upon which the assumptions rest.

We then move backward to the Search and Discovery and Information Management elements. Here, we might examine the following: for human intuition or reasoning that is applied in the Predictive Analytics element, could the intuition or reasoning be encoded so that it is available and searchable for others? And can the intuition somehow be validated? For computer-based predictive analytics such as price-elasticity models and competitor bid-pattern models, we might examine the information that feeds the models, the quality and timeliness of the information, and so forth.

And finally, we examine the Data Acquisition and Data Filtering elements of the data-to-learning-to-action process for the decision. Here, for human intuition-based decisions, we might explore what data the intuition is based on. Do we have the right *sources* of the data that are needed to attain the insights sought? If it is historical data, is that data still valid for the *current environment*? Given the well-known human data-filtering heuristics (e.g., confirmation bias), we would want to understand if the historical data might be susceptible to being misinterpreted or distorted due to cognitive biases such as selection bias. For the information that feeds model-based predictive analytics, we would want to know the quantity, quality, and timeliness of the data that underpins the information and how it was implicitly or explicitly filtered. By the time we complete our first-cut understanding of the Data Acquisition element, we will have a good overall understanding of the data-to-learning-to-action process for the pricing decision, and we can move on to a second, more systematic pass, which we will do after briefly working through a couple more first-pass examples.

As a second first-pass example, let's assume that a value driver for a technology company is the acquisition of computer programming talent. It's a value driver because the company has found that the difference in the quantity of programming output and the quality of the output can vary by upward of an order of magnitude among programmers, and programming output and quality ultimately have a significant impact on the company's financial performance. The decisions we therefore want to improve are the decisions on acquiring programming talent.

Working backward from the decision, we first examine the Process and Collaboration element of the data-to-learning-to-action process and determine who is making the acquisition decision and how the decision is being made. Are there discernible patterns in which some people, individually or in groups, are making better programmer-acquisition decisions than others? Are some processes for making the decisions seemingly more effective than others? For example, with more people inputting into the hiring decisions, are the outcomes better, or are they actually worse?

Then, moving back to the Predictive Analytics element, we want to understand what types of models or heuristics are applied to try to predict the performance of candidate programmers. Is it primarily human intuition, or are there models that take as inputs various sources of information such as résumés and interview results and use that information to make predictions? We might ask what information, if available, would be useful to include in the predictive models and why.

Moving backward to the Search and Discovery and Information Management elements, we might ask about the accessibility of résumé information by the decision makers. We might also want to know if historical interview information has been captured and stored so that it can be analyzed within the context of the Predictive Analytics element. If there was beneficial information identified in the Predictive Analytics element that was not currently included in the predictive models, we would want to know if it was already available within the Information Management element.

Then, at the Data Acquisition and Data Filtering elements, we might want to know if candidate interview information could be captured, if it was not already, and how exactly we would do so. If there was information that was identified by the Predictive Analytics element that was not currently in the model, say, sample programming code, that could not already be attained from the Information Management and Search and Discovery elements, then we would want to understand if the data could be acquired and/or filtered, and at what cost.

Let's look at one last brief first-pass example, in which the value driver is the number of patented inventions generated by the R&D organization, and the primary decision associated with this value driver relates to the question of what ideas should be patented versus remaining as trade secrets. Again, we would first look at the Process and Collaboration element to determine how the decision is made. We would probably find that the decision is made by a committee that periodically considers candidate invention disclosures. We would want to know how the decisions are made—for example, who is involved, and does majority rule or is unanimity required? Are the decisions made in live meetings or off-line? What other information would be useful to the decision participants that they don't already have when making the decisions?

We then look to the Predictive Analytics element and determine if there are predictive models that provide guidance on whether to pursue an idea as a patent. These predictive models could reside within minds or machines, or both. The models might consider product directions (e.g., "the subject matter is not a long-term direction for us, so we won't spend money on patents in that area"), and/or they might include such factors as the expected licensing value of the patents. As is typically the case for this element, we would want to understand what information could be useful to include in the predictive models but that currently is not. For example, detailed information on the probability of success in attaining a patent for specific technology classes in specific countries might be important.

Moving backward to Information Management and Search and Discovery, if there is information that decision makers would like to have but do not currently have, and/or that would be useful to include in predictive models but that is not currently included, we would want to know if that information was already in the Information Management element. If so, we would want to explore the possibilities of it being made available to the models or directly to the decision makers (in which case, it is important that we take the perspective that the modeling is indeed occurring—it simply happens to be purely mental in nature).

And finally, if there was useful information for predictive modeling and decision-making purposes that was not available in the Information Management element, for example, historical data on the probability of success in attaining a patent for specific technology classes in specific countries, we would want to explore the possibilities of that data being acquired and the expected cost of doing so.

These have been just a few examples of the myriad decisions that are made by organizations that can be examined in the context of the data-to-learning-to-action process. And as mentioned earlier, even decisions that relate to areas that are not traditionally thought of as being connected to financial statements, such as employee morale, can be worked the same way, with the caveat being that they must be explicitly linked to, and have significant leverage on, financial outcomes, which, of course, is a prerequisite for being a candidate value driver.

Working backward in this manner provides a *basic* understanding of the *flow of learning* that contributes to the targeted decisions. It is a useful first pass that helps in getting a "lay of the land" as well as with building a rapport with participants in the data-to-learning-to-action process. But notice we don't yet have much information on what the *value* of the flow of learning is for the data-to-learning-to-action process we are analyzing or the *limiting factors* to the flow. We may have a general sense and some working hypotheses for that at this point, but a general sense alone can lead us astray, so we have more work to do.

Constraints on Value

Let's return to our pricing decision example. Now that we have a basic understanding of its data-to-learning-to-action process from the first pass, we will make another pass working backward from the pricing decision, but this time we will do so in a more structured way that will reveal the limiting factors in the process.

We found in our first pass that at the Process and Collaborate stage of the chain the pricing decision is made by the consensus of several people, including the sales lead and management, and this consensus is reached through communications over email. We now want to understand two fundamental things with respect to this element of the process: 1) to what degree, in terms of value, do *activities and operations within* the Process and Collaborate element contribute to the failure to close profitable deals, and what would be the *expected value* of resolving these *internal* constraints, and 2) what *inputs from the upstream elements* of the chain (e.g., Predictive Analytics) are the activities and operations of the Process and Collaborate element dependent on, and what would be the *expected value* of improving these inputs?

For the first question, relating to the value constraints attributable to the internal operations of the Process and Collaborate element, we might find, for example, that occasionally deals have been lost that otherwise were expected to be won because of the amount of time required to come to a consensus on a decision due to the going-back-and-forth via emails among, say, four parties. And if that problem is not *dependent at all* on any inputs from upstream elements, we can immediately conclude that it is a *limiting constraint* on value.

We next want to work toward calculating the cost associated with this limiting constraint, although the rub is that the deals-that-have-been-lost-that-otherwise-were-expected-to-be-won factor will probably require some subjective judgements by those most knowledgeable about the situation. But that's perfectly fine—it is certainly better than failing to quantify the issue at all! That's because *failing to quantify* generally means *failing to do anything* about it, or doing the *wrong* thing, because without quantification there are only anecdotes, and anecdotes are simply not sufficient. Throughout the approach outlined in this book we strive to quantify when at all possible for just that reason, but very often those quantifications necessarily rest on insights that reside in people's heads. And the good news is that there is invariably a significant amount of those insights that have never been fully tapped!

So, with the cost of the limiting constraint that is related to the *internal operations* of the Process and Collaboration element estimated, we turn to the second question, which is related to the inputs from the upstream elements, such as the Predictive Analytics element, that the Process and Collaborate element is dependent on. Specifically, what we want to understand is the value of receiving better or different inputs. (And we also would want to understand if any inputs from upstream elements that *are* currently flowing to the Process and Collaborate are not needed by the element and therefore deliver *no value* to the element.) So, we ask of the participants of the Process and Collaborate element: "What information that you don't currently have would enable you to close more deals?" If we receive an answer that there is indeed information that would enable more deals to be closed, we know immediately that there is another limiting constraint, this one occurring somewhere upstream of the Process and Collaborate element.

We would then follow up to understand the cost of this second limiting constraint on value: "What would be the expected value of those additional closed deals if you had that information?" Again, when we pose these questions we are looking for subjective estimates. We may be able to then build computer-based value models that are based upon these subjective estimates, but the estimates are the foundation. To make the foundation as strong as possible we must ask the right questions of the right people. We will cover how to do that in more detail in later chapters—here we are just concerned with walking through a general example.

Let's assume that the Process and Collaborate participants indicate that understanding competitor bidding better is the most important information enhancement. While we learn that a better understanding of the price elasticity of the customer would also be helpful, it turns out that most often deals that are lost are lost to competitors, so understanding their likely bids is the dominant value-constraining factor. We would further want to get a sense of the value of being able to better predict competitor bids at various levels of accuracy, ranging from perfect insight (probably unattainable, at least legally!) to only slightly better than current. Again, these are most likely subjective estimates, but now we have something to work with.

Now, as is always the case with the Predictive Analytics element, for this pricing example the operations of the element may be embodied within the minds of the participants in the Process and Collaboration element or within the minds of others in the organization, and/or they may be embodied within computer-based modeling. But whether explicit or implicit, predictive models certainly exist, and we want to know exactly what learning flows within the data-to-learning-to-action chain they are dependent on. Let's assume there is at least a simple computer-based model that predicts competitor bidding, but per our feedback from the consumers of the output of this model, the decision makers in the Process and Collaboration element, it is not very accurate. We therefore ask those in charge of the model: "What information would enable the model to be more accurate, and how much more accurate?" They answer that understanding the variable costs associated with the competitor's solution is the most important information since the competitor will not bid below its variable costs.

We now turn to the Search and Discovery and Information Management elements. Is information on competitor variable costs available within the organization, and can it potentially be accessible by the competitive-bid modelers? If so, we have identified where our other limiting constraint on value for the data-to-learning-to-action process that is associated with the pricing decision resides. The information exists; it is just that something related to the Search and Discovery element and/or the Information Management element is causing the constraint.

But let's assume that the information does not already exist anywhere within the organization. We turn to the Data Acquisition and Filtering elements to determine if there are ways to obtain the required data. This is where innovative thinking is often needed because it is likely that no one has ever thought about acquiring such data (because no one ever did the type of analysis we just conducted!). Or perhaps it was looked at before but was never resolved because there was no quantification of the benefits of having such data, a particularly common situation if the only relevant data that could be attained was going to be less than perfect.

So, the next question we ask is: "Can we acquire competitor variable-cost data?" No. "Ok, are there other data that *could be* acquired that would allow us to at least *estimate* variable costs?" We find that the answer is that competitor variable cost is primarily determined by the cost of the materials that are included in their solution, and these materials costs can be estimated. In other words, we have found a way to reduce the uncertainty we have with respect to competitor variable costs by understanding more about competitor materials costs.

We have therefore identified the other limiting constraint of the process—uncertainty with respect to competitor variable costs. And we have determined a solution to help resolve some of that uncertainty: if we could attain data that enables us to estimate reasonably well a competitor's cost of materials for a given solution we could estimate the associated competitor's variable costs. That would enable better pricing decisions and thereby significantly increase the value of our data-to-learning-to-action process, and this increased value would inevitably appear directly in the financial results.

Often there will not be a *perfect* solution to the constraining factor, at least for the near term. But even a little improvement can go a long way. A key point here is that when there is a chain of direct dependencies in a data-to-learning-to-action process even *a small improvement to the limiting constraint is worth more than big improvements to other constraints!* This is perhaps the key point of the book because this rule is so often violated in practice. It is *better to be approximately right in addressing those things that are truly most important than to be precisely wrong by getting all the details right on what is not most relevant to the ultimate performance goals.*[3]

Even a small improvement to a *limiting* value constraint is worth more than large improvements to other constraints.

On the other hand, sometimes there are independent learning flows and associated constraints, as was the case in this example, that can be worked on in parallel if resources permit. For instance, the smaller email-based limiting constraint could be targeted in addition to the higher-leverage competitor bid–prediction opportunity since there is not a direct dependency. However, the important caveat is, "if resources permit." Too often opportunities for

---

[3]An important sentiment that pervades this book. Often attributed to the economist John Maynard Keynes, but apparently first articulated by the philosopher and logician Carveth Read: "It is better to be vaguely right than exactly wrong". Carveth Read, *Logic, Deductive and Inclusive* (1898), p. 351.

improvement with much less upside are targeted because they are simply easier to identify and address. And whether the target is big or small, the quantification of the expected value of different solutions aimed at addressing the constraint, as well as a corresponding quantification of the expected costs of each of the potential solutions, is an imperative.

As we worked our way backward from the decision being addressed, we asked the following fundamental questions for each element of the data-to-learning-to-action process:

- What *internal* activities and operations within the element that we are currently examining are constraining value? What would be the value of resolving the internal constraints?

- What *inputs (or lack thereof) from upstream element(s)* are constraining the value of the element that we are currently examining? What would be the value of resolving the input constraints?

Working backward from decisions and asking these fundamental questions is the universal procedure for determining the limiting constraints on value, as is illustrated by Figure 5-5, which we can apply to *any* data-to-learning-to-action process that we want to optimize.

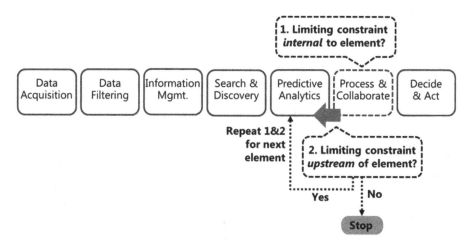

Figure 5-5. General procedure for determining limiting constraints on value

Resolving Constraints

In our example, we identified a limiting constraint on value as being due to too much uncertainty about competitor variable costs, which causes uncertainty about competitor bids, which inhibits the optimal setting of prices for our own solutions for customers, thereby causing us to lose what could otherwise be valuable deals (or to under-bid). And we are uncertain about competitor variable costs because we are uncertain about the primary component of the variable costs—the costs of materials in competitors' solutions. So, if we can reduce uncertainty about the costs of materials in competitors' solutions, we have learned something that will ultimately have a positive impact on our value driver.

In some cases, resolving a constraint can be more straightforward than expected because the constraint was simply never identified as such before, and so no one seriously addressed it. This happens more often than one might think, because without systematically working through our approach here, important constraints can simply be overlooked. Resolving constraints on learning and value takes organizational energy and creativity, and if the cost of the constraint isn't quantified, those creative resources are not likely to be allocated. Another reason even obvious constraints may not have been addressed is because of the curse of perfect accuracy: if we can't have perfect predictive accuracy, we don't bother to get *any* extra accuracy at all because we just *ignore* the uncertain variable. It's a throwing-the-baby-out-with-the-bath-water syndrome.

In our example case, perhaps interviews of customers who purchased the competitor's solutions could enable us to better understand the brand of products the competitor used in its solution, and when combined with price quotes from the supplier of the brands, would enable a decent estimate of the materials costs of the competitor's solutions. This data may be far from perfect, but it can still be expected to reduce our uncertainty to some degree about competitor variable costs, and that will likely make a difference.

There may be a range of solutions for resolving any given constraint in a flow of learning, each with a different level of expected efficacy in reducing the uncertainty that is the cause of the constraint, and therefore each delivering a different level of value. Each of these solutions will require some degree of change in people, process, or technology from the current state, and each will also have an expected cost to implement. How do we decide which of the solutions to implement? As a first approximation, the same way as for any other type of potential investment an organization makes—typically, by choosing the solution with the greatest net present value of the expected cash flows associated with each solution. I say as a "first approximation" because, as we will discuss later, if we do a look-ahead constraint analysis, we may find it would make sense to resolve the constraint more (reduce more uncertainty) than would be warranted by applying a purely one-constraint-at-a-time approach.

The approach described in this chapter alone will deliver significant benefits to just about any organization. But we can go further by tightening up both the estimations of the foregone value due to limiting constraints as well as the estimations of the value of implementing solutions to at least partially resolve the constraints. We'll delve into detailed methods for quantifying these estimates in the next chapter.

Summary

In this chapter, we discussed how to decide which of the myriad decisions in an organization should be the focus of our attention and resources, with value drivers being the proper guide for these prioritization decisions. Once a high-leverage decision is identified, we work backward from the decision along the data-to-learning-to-action chain to identify value constraints. At each element of the data-to-learning-to-action chain we determine if the limiting constraint (or constraints, if they are independent) on value is due to a deficiency within the operations of the element itself or if the limiting constraint is due to a deficiency of an input from an upstream element. We continue this process of working backward along the chain until we have identified all the limiting constraints. We also explore the possibilities for resolving the identified limiting constraints and strive for an understanding of the associated expected value of each of the possibilities.

Quantifying the Value

In the last chapter, we reviewed some examples of working backward from decisions to understand the value of addressing bottlenecks on actionable learning. For instance, for the pricing example, we needed to understand how much more profit would be expected to be attained if we could better estimate competitors' bids. And, more particularly, we wanted to know how much more profit would be expected for *various levels of accuracy* in predicting competitors' bids. With that information in hand, we could then work backward in our straightforward way to identify the constraints that contributed to the current state of less-than-perfect-predictability of competitors' bids and determine what it would be worth to resolve those constraints, starting with the constraint that was identified to be the current most-limiting factor in the data-to-learning-to-action chain.

But we didn't detail exactly *how* we could develop estimates of how much more profit would be expected for various levels of accuracy in predicting competitors' bids. In this chapter, we will review ways to do that so that an *expected learning value* can be estimated for debottlenecking a data-to-learning-to-action process. Doing so provides a quantification that can serve as a basis for deciding on *investments in learning* that can compete effectively with the other demands for investments and resources across an organization.

This is the main chapter in the book in which the quantifications that we will discuss amount to more than basic arithmetic. Even so, we are not going to work through detailed calculations, but just provide an overview of the

© Steven Flinn 2018
S. Flinn, *Optimizing Data-to-Learning-to-Action*,
https://doi.org/10.1007/978-1-4842-3531-7_6

process of how to perform them. If you want to have a general understanding of how to quantify learning value but leave the modeling and calculation details to others in your organization, just skim through the chapter and focus more on the "Learning Value Modeling" section and the summary at the end of the chapter. You will have all the background you need for the rest of the book.

Value of Learning

The value of learning derives from the learning's potential to affect one or more decisions. As we discussed in earlier chapters, this is a derivative of the concept of value of information in the field of decision analysis, which is also sometimes known as the *value of clairvoyance*—that is, it is the value of being able to attain additional *foresight* about what the outcome of an uncertain variable will be that can potentially serve to change a decision from what the decision would otherwise be. Now, our hypothetical clairvoyant can be of the infallible type, in which case we would say that the value of the information provided to us is the *value of perfect information*, or, if our clairvoyant sometimes predicts incorrectly, then the value of the information provided is the *value of imperfect information*, which, of course, will necessarily be worth less (or at least no more) than perfect information.[1]

The reality is that *most* information that is available in the real world provides a less-than-perfect basis for predicting uncertain outcomes, and further, the *information itself* only provides a *basis* for what really counts, which is the *conversion of the informational inputs* into *improved predictions* of the outcomes of uncertain variables. That conversion requires some sort of modeling—or, most generally, learning—capability, whether occurring in minds or machines. In classic decision analysis, this learning is implicitly embodied within the specific analysis that is performed for a particular decision, analysis that is often performed by third-party experts, and hence the label "value of information" tends to be suggestive of this paradigm of larger, one-off decisions, with associated specialized expertise and techniques applied.

Most generally, however, that type of specialized analysis—as well as any other type of process, whether performed by minds or machines, that results in the adjustments of representations of uncertainty based on informational inputs—is a form of *learning*. It is the result of this learning that has potential value, not the raw informational inputs themselves. And, of course, this value is amplified when the conversion of information to predictive insights is an ongoing process that is institutionalized within an organization.

---

[1] For a comprehensive review of decision analysis, including value of information–related concepts and more-complex examples than those discussed in this chapter, see Howard and Abbas, *Foundations of Decision Analysis*, Pearson Education Ltd (2016).

Information *by itself* does not have value. It is the *conversion* of the information into *improved predictions* that can affect decisions (i.e., actionable learning) that has value.

Hence, for our purposes, the value of learning is the more meaningful term to use, rather than the value of information. And yes, you may have noticed that this implies that there is necessarily a data-to-learning-to-action process at work in calculating the value of learning for a data-to-learning-to-action process, so it is turtles all the way down![2]

Before we get into the details of value of learning quantifications, it is useful to review some of the fundamental tenets and insights from the field of decision analysis that are universally applicable. First, for any specific decision, we need to understand the potential actions that can be decided upon. For recurring decisions that have been previously made, this can be straightforward as the potential choices have likely been well established. But valuable innovation is often the result of expansively considering new options not previously considered. Creatively determining additional options is therefore an important aspect of optimizing data-to-learning-to-action processes. The good news is that because we have a more structured and effective way to assess *the value* of alternative actions, it behooves us to consider as many options as possible, even if some of them may feel like a "stretch." Although this means that many will be discarded when more thoroughly analyzed, the one or relatively few that make it through the value filters may make all the difference.

And the analysis of the possible actions that are *not* chosen may provide insights about under what conditions the discarded action might be preferred because it could potentially generate additional value for the organization versus the current decision choice. For example, when an applicable technology sufficiently matures, perhaps an action that is dependent upon that technology would become preferred. Identification of those events and conditions that would potentially change chosen actions provides a basis for then tracking the event or condition of interest over time.

Second, because the focus of our approach is with respect to businesses, the value of the outcomes of actions can generally be assessed directly in monetary terms, which avoids the need to compare "apples and oranges" for decision outcomes. Furthermore, while generally decision analysis can address situations in which there is a risk-aversion overlay on the various outcomes of a decision, here we can assume risk neutrality, which is realistic for most corporate decisions, and simplifies quantifications. In other words,

---

[2]http://lithub.com/turtles-all-the-way-down-stephen-hawkings-a-brief-history-of-time/

if an expected outcome of a first action is worth $1 million and the expected outcome of second action is worth $10 million, the second action is worth exactly ten times as much, or in decision analysis terms, the *utility* to the decision maker is exactly ten times as much. And the decrease in utility of losing $10 million is of the same magnitude as the increase in utility of gaining $10 million. In atypical situations in which relaxing the assumption of risk neutrality is appropriate, the quantification of the value of learning is a bit more complicated (requiring the application of explicit utility functions), but the overall procedure is the same as the risk-neutral approach that we will concentrate on here.

Third, a key maxim of decision analysis that holds for our approach as well is that the quality of a decision can't be judged by its outcome. This follows from the fact that since uncertainties affect the outcomes of decisions, good decisions can always run afoul of bad luck. Decision quality must be judged on an *a priori* basis. That is, given the informational inputs at the time of the decision, or, most broadly, the *state of information*, is the decision that is taken the best one? That's the question of *decision quality* posed by classic decision analysis. However, decision quality is a function of that necessary intermediary between information and decisions—learning. The more germane question then, is: "Given the information available at the time of the decision, is the *learning* that is based on the information and that informs one or more decisions as good as it can be?" In other words, decision quality is really a subset of *learning quality*.

The quality of a decision cannot be judged based upon the *outcome* of the decision. It is the associated *learning* given the state of information *at the time of the decision* that should be evaluated.

Calculating Learning Value

Let's now move on to the process for quantifying the value of learning by returning to our bid-pricing example. For that example, we found that a constraining factor on value is the uncertainty with respect to competitors' bid prices, and therefore we want to explore solutions for reducing that uncertainty. So, we want to understand what the expected value would be for different levels of predictive accuracy on our competitors' bid prices so that we can then compare those values with the costs of solutions that provide the associated levels of predictive accuracy.

There are fundamentally two outcomes that can result from our bid that we need to consider. The first situation is one in which we lose the deal because our bid is *higher* than a competitor's bid (or if there are multiple competitive bids, our bid is higher than the lowest of the competitors' bids), as depicted in Figure 6-1. In such cases, our loss of profit compared to what it would have been if we had perfect predictability of the competitor's bid is the difference between the competitor's bid and our breakeven cost for the project that we are bidding on. In other words, if we could have the benefit of perfect predictive capabilities with respect to a competitor's bid, we would simply price our solution just a tiny bit below the competitor's price, and if there were no factors other than price that affected the customer's decision, we would win the deal.

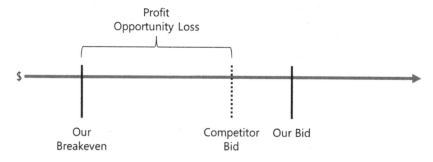

Figure 6-1. The case in which we lose the deal

The other case is the situation in which we win the deal because our bid is *lower* than the competitor's bid, as depicted in Figure 6-2. In that case, the good news is that we will have an expected profit that is the difference between our bid and our breakeven amount. The bad news is that compared to the situation in which we have perfect foresight of the competitor's bid, we leave money on the table—which is the difference between the competitor's bid and our (winning) bid.

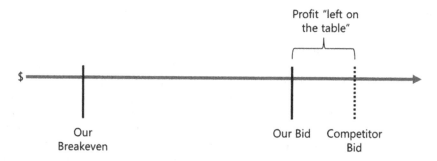

Figure 6-2. The case in which we win the deal

So, fundamentally, what we want to do is minimize the profit left on the table when we win bids because we bid too low relative to the competitor's bid, while at the same time minimizing the probability of losing deals and the associated profit opportunities because we bid too high relative to the competitor's bid. Clearly there is tension between these two objectives, and the more uncertain we are about the competitor's bids, the greater our expected foregone profits are versus the case in which we had perfect predictability of the competitor's bid.

Encoding Uncertainties

Based perhaps on historical competitor bids, we have a current view of the uncertainty of competitor bid prices, roughly represented by the probability function depicted in Figure 6-3. This probability function looks somewhat bell shaped—in practice it could have a significantly different shape, the precise contours of the function being dependent on what is already known about competitor bidding. For example, if we knew very little about competitor bidding, the probability function could be a uniform probability function that would be represented by a horizontal line stretching across a large range of price bids. This probability function would imply that we believe that *any competitor bid is as likely as any other* within the range. The bell shape of Figure 6-3 implies that we are not *completely* unsure about competitor bids, however. The bell-shaped probability function implies that it is more likely that a competitor bid price will be at a level that is near the center of the bell than at the tails. Still, there is a significant spread to the bell, embodying our current significant uncertainty about the competitor's bid.

Probability of Competitor Bid Price Levels ($)

Figure 6-3. Current probabilistic assessment of competitor price bids

How do we create a current probability assessment such as the one depicted in Figure 6-3? Often, we do so by encoding the expertise on the subject from the people who are most knowledgeable about it. That can be done by conducting structured discussions with them. Alternatively, the probability function of Figure 6-3 might be generated directly from computer-based modeling that, for example, analyzes historical competitor bid patterns and generates the probability function based on that analysis. Of course, this type of modeling is an area in which data scientists can make a significant contribution, and the tools that are at their disposal keep getting more effective at being able to translate historical data into probability functions.

Regardless of how probabilities are generated, an important point about probabilities needs to be emphasized: events and objects do not have an *intrinsic* probability; rather, a probability is an *extrinsic assessment* that is based upon a particular *state of information and associated learning*. That fundamental aspect of probabilities is always true, whether a probability is assessed directly by a person or generated by computer-based modeling.

A probability is an *assessment* of uncertainty that is based on a state of information and learning, not an *intrinsic quality* of an event or object.

If encoded by interviewing human experts, the uncertainty is typically encoded as a *cumulative probability distribution*, as illustrated in Figure 6-4. This format tends to be more amenable to directly encoding and cross-calibrating interviewees' answers to questions such as "What is your assessment of the chance that the competitor's bid will be above price A?" and "What do you think the chances are that the competitor would price below price B?" The answers to a series of these types of questions results in a probability distribution function as depicted in Figure 6-4; specifically, a cumulative function. This cumulative probability function appears to be continuous rather than comprising just discrete points, which might be a result of interpolating between points provided from our interviews. Regardless, the same basic procedures outlined in this chapter can apply to either continuous or discrete probability distributions.

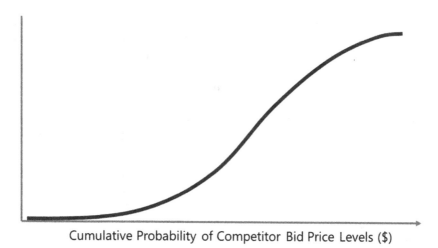

Cumulative Probability of Competitor Bid Price Levels ($)

Figure 6-4. Current cumulative probability distribution of competitor price bids

The function depicted by Figure 6-3 is an alternative representation of the same probabilistic assessment, formally called the *probability density function*. The density function is, in calculus terms, the derivative of the cumulative distribution function; or, reversing the perspective, integrating or cumulating under the density function yields the distribution function. While the cumulative distribution function is generally easier to use when encoding expert knowledge, the density function usually makes it easier to see at a glance the differences in the profiles of uncertainty, and so it will be more typically used to illustrate uncertain variables here.

Now, perhaps the method of extracting probability distributions from those with relevant expertise on the subject strikes you as being quite "subjective." Perhaps you would be more comfortable if the probabilities were derived by the application of, say, a complex neural network. The good news is that they are! *Subjective* is just another word that we use for the output of neural network–based models. They just happen to be models operating on a biological-based substrate! And this is not meant to be a facetious comment; it is a key point in the optimizing data-to-learning-to-action approach. A significant amount of value is lost by organizations because *subjective* is a dirty word to some people. Rather than capturing relevant subjective insights by carefully determining and encoding them through structured techniques, they are often overlooked or ignored. The result is that valuable insights that should otherwise be available to the organization are not applied, resulting in poorer-performing data-to-learning-action processes than would otherwise be the case.

There is a psychology that many people succumb to, and not just engineering types, that if numbers come out of a machine they are somehow more credible than if they come out of a mind. That may *sometimes* be the case, but it may also be exactly the opposite. The human mind's being approximately right on a topic, for example, is much more valuable than a computer model that is precisely wrong because the model fails to account for all the relevant complexities of a situation. The human mind certainly has its known quirks and weaknesses with respect to certain classes of uncertainties, but it also tends to be quite good at getting it pretty close to right in many highly complex situations—situations that automated systems continue to struggle with. So, always, the key is to find the right combination of mind and machine-based learning that provides the best learning and predictive results for a given situation.

Now, we can only very briefly cover the mechanics, and art, of encoding uncertainties in this chapter, whether the source is mind or machine. For encoding from minds, we touch on basic interviewing techniques and cross-calibrating among interviewees. There are many other tools and techniques that can augment these basics. For example, *systems thinking* techniques can be helpful in contemplating the uncertainties related to the dynamics of systems that have large numbers of interacting parts. Applying personas and empathy maps that help you to understand different perspectives may be useful in considering uncertainties that are associated with human behaviors. Scenario analysis may be helpful as well. And beyond these techniques, it is important to explicitly work to overcome the myriad biases that people are prone to, such as selection biases in recollecting historical information, overconfidence in the ability to predict, and groupthink. For example, "Black swan" events may be underappreciated, perhaps at least in part due to our general overconfidence in our ability to accurately predict.[3] On the other hand, selection bias and emotional resonance may work to make unusual events more memorable and hence unduly influential in probabilistic assessments compared to more mundane events. Similarly, storytelling is currently a popular technique that resonates with many people and can help them learn and retain information. But the power of stories may sometimes be exploiting a mental bug rather than a feature, because stories are necessarily selective anecdotes and may not be representative of the reality that we want to underpin our encoded uncertainties. So due care is required in applying any specific method to facilitate quantifying uncertainties.

It is also clearly important to *ask the right questions* to get to the best possible uncertainty coding, which is a combination of both science and art. In that regard, as we will discuss in an upcoming chapter, it can be useful to have an interviewer who does not necessarily have significant knowledge of the subject matter being queried so that the "naïve" questions are asked that can elicit a broader range of perspectives than might otherwise be surfaced.

But regardless of the techniques that are applied to attain the best possible understanding of the uncertainties of interest, it is imperative that we encode uncertainties into explicit probability distributions. The relevant question should be *what* the probability distribution should look like based on all the various considerations, *not* whether to bother to encode uncertainties into probability distributions at all. So, for example, taking into consideration Black Swan–type thinking, we might fatten the tails of a distribution, everything else being equal, rather than simply throwing up our hands. On the other hand, ensuring that selection bias of unusual events doesn't unduly influence a probabilistic assessment may prompt us to add more mass in the center of a distribution than would be the case without explicitly considering the selection-bias effect.

---

[3]Taleb, Nassim Nicholas. *The Black Swan: the impact of the highly improbable* (2nd ed.) (London: Penguin, 2010).

Encoding uncertainties as probability distributions should always be a matter of *how* to do it, not a matter *whether* to do it!

By performing the encoding of uncertainties into probability distributions we are simply making explicit and more rigorous what would occur by default in any case, and it enables us to effectively quantify the value of learning. It also serves to enforce a mental discipline that can promote clearer thinking on the factors that contribute to the uncertainties being considered.

Value of Learning with Perfect Predictability

Given the current level of uncertainty that is embodied by the probability functions of Figures 6-3 and 6-4, we can determine the bid that can be expected to optimize our profit. Typically, we would do that by application of a computer-based *simulation* because, in general, our probability functions may not be amendable to deriving "closed form" solutions of the type that are often found in text-book examples, and this is even more likely to be the case as we add more modeling details, such as additional uncertain variables. In the simulation, we select a series of hypothetical bid prices that we make and then for each of these bids perform a sufficiently large number of samplings from the probability function represented by Figure 6-3 to simulate the competitor bids. The expected profit is recorded for each instance of our simulated bid and corresponding competitor bid. Our simulated bid that averages the greatest profit represents the price that we should use for our real-world bid. With sufficient simulation runs and granularity, we can achieve an arbitrarily high level of confidence in finding our optimal bid, along with the expected profit associated with this bid.

With the optimal bid calculated given our current assessment of competitor-bid uncertainty, we can next calculate what the value would be to have *perfect predictability* of the competitor's bid. That is simply a matter of comparing the expected profit that we would have with perfect foresight versus the expected profit given the current state of uncertainty that we just calculated with our simulation.

The expected value of learning is always equal to the value that is expected *after* the learning has occurred *minus* the value that is otherwise expected without benefit of the learning.

Of course, in this simple example, with perfect predictability of the competitor's bid, we know that we would always bid just slightly below the competitor's predicted bid price (so long as it was above our breakeven point). But how do we know *what* the clairvoyant will predict? Well, the clairvoyant can only *predict* the future, not *change* the future (a hypothetical entity that is able to change the future, a much greater power than clairvoyance, is termed a *wizard* in decision-analysis parlance). The clairvoyant's predictions will have to obey the only view of the future that we currently have, the probability function of Figure 6-3. That is, the clairvoyant's predictions are necessarily *conditional* on our current probabilistic assessments. So, we can model what the clairvoyant is expected to predict by discretizing the competitor bid price probability function into discrete slices and then sequentially selecting each possible bid-price level and its associated probability of occurring, as illustrated by Figure 6-5. This represents the range of predictions and the associated likelihood (which is the area under the curve of each discrete slice) of each prediction that the clairvoyant can be expected to make.

Figure 6-5. Sketch of the process for determining the value of learning with perfect predictability

So, again, there is still *a priori* uncertainty because we have a clairvoyant, not a wizard, helping us. But in this case, we (or our model) make the bid-price decision *after* the prediction of the competitor's bid is received, and the prediction is *guaranteed to be accurate*. So, it is optimal to always just bid a bit below the competitor bid that the clairvoyant predicts, as is shown for a given prediction slice, *k*, in Figure 6-5. Unlike the original case, with perfect predictability our optimal bids now always win the deal and deliver a profit, although in some instances the profit may be relatively small because the predicted competitor bid is low. The total expected value of the learning in this case of perfect predictability (where the "learning" here is having the services of the clairvoyant) is simply the sum of all these individual profits weighted by the probability function of the competitor's bid price, which the clairvoyant's predictions necessarily obey, minus the expected profit from the original case. Or, in equation form, if we have *n* prediction slices:

Expected Profit After Learning

$$= \sum_{i=1}^{n} Expected\ profit\ optimal\ bid\ slice(i) * Probability\ slice(i)$$

Expected Learning Value
= Expected profit after learning
− Expected profit before learning

While an assumption of perfect predictability may be highly unrealistic, it provides an *upper bound* on the value of any possible combination of additional information and analytic techniques (i.e., learning) that are targeted to improve the predictability of competitor bids. That can be very useful in performing sanity checks that can save time and money that would otherwise be lost attempting to pursue solutions in which the cost would be greater than any possible benefits.

Calculating an expected value of learning assuming *perfect* predictability of the uncertain variable is typically easier than doing so for *imperfect* predictability and provides a *useful screen* for filtering out solutions that would cost more than any possible benefits they could deliver.

Value of Learning with Imperfect Predictability

The more likely real-world situation is one in which there may be opportunities to improve predictability to some degree, but far from perfectly. In our bidding case, as we saw in the last chapter, understanding competitor materials costs would enable an improved understanding of competitor variable costs, and therefore would result in better predictability of customer bids than currently. In other words, the ability to model competitor materials costs should enable a shrinking of the probability density function from that seen in Figure 6-3.

So, how can we get to that shrinking of the current probability function based on the competitor materials costs learning? We can develop a new predictive probability distribution that is *conditional* on our prior distribution, such as is illustrated by the dashed function in Figure 6-6. This distribution represents our updated uncertainty with respect to our predictions of the competitor's bid price, a reduced level of uncertainty from what we originally had because of our analysis of competitor materials costs. We might expect this new probability function to be asymmetric, as shown in the figure, because our modeling of competitor variable costs would likely provide tighter lower bounds on the competitor's bids than upper bounds.

And how do we get this probability function of bid-price predictions that is based on our attaining an enhanced understanding of competitor materials costs? Probably the same way that we got the original bell-shaped probability distribution function. Quite typically, we will encode such probabilistic assessments from human experts, perhaps in combination with the application of computer-based modeling. We perform this encoding by asking a series of questions of people with the most relevant expertise and experience to understand the shape of the probability distributions that embody the uncertainties. Again, this is generally done by encoding the expert views as cumulative probability distributions (as depicted in Figure 6-4) as opposed to a probability density function form (as depicted in Figure 6-6). If there are multiple human experts with differing opinions, we would employ procedures to try to reach a consensus. This consensus could be achieved through dialog with the experts to determine if there are differing assumptions, bases of knowledge, and so forth that, if shared, would help move toward a consensus. Alternatively, in accordance with wisdom-of-the-crowd-based thinking, the expert views might simply be averaged.

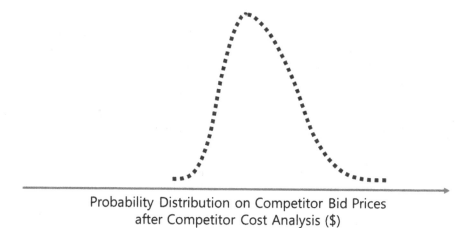

Probability Distribution on Competitor Bid Prices
after Competitor Cost Analysis ($)

Figure 6-6. Conditional predictive distribution on competitor bids

So, the next question is, "What is the *expected value of the learning* that is represented by the after-competitor-cost-analysis probability function that is depicted in Figure 6-6?" We already know the expected value of perfect predictability of competitor bids that we worked through in the previous section, which provides an upper bound on the value that we can attain with less-than-perfect foresight. Perfect foresight would be the equivalent of the after-competitor-cost-analysis probability function shrinking to just a vertical line in Figure 6-6. And given such perfect foresight, we know that the obvious optimal bid is to set our price just below the predicted price of the competitor. But merely understanding more about competitor materials costs clearly leaves us well short of that perfect foresight, as we can see by simple inspection of Figure 6-6.

So, let's now walk through how to determine the learning value associated with this less-than-perfect-but-better-than-we-had predictability of the competitor's bid. As in the case of perfect predictability, we calculate the learning value by comparing the expected value of the deal that is based on our current level of uncertainty of the competitor's bid with the expected value of the deal if we went ahead and performed the materials cost analysis that would enable us to shrink our uncertainty to some degree.

We already discussed the mechanics of how we can determine the expected current value of the deal that we plan to bid on (most probably via a computer-based simulation). We also discussed how to determine the expected profit of perfect predictability of the competitor's bid. In that case, once we received a prediction (a prediction in accordance with our current state of learning, the original bell curve of Figure 6-3) we simply bid just slightly below the competitor's bid. We were therefore guaranteed to always win the deal and attain as much profit as possible by simply bidding just below the competitor's bid.

But this case is a bit different. With perfect foresight, we could just apply a straightforward rule after receiving the prediction—bid just below the predicted competitor bid. Here, we instead have uncertainty *even after* we receive the prediction, albeit less uncertainty than we had before receiving the prediction. We can use the same basic procedure as in the case of perfect foresight, but with an additional simulation required that addresses the *residual uncertainty* that exists even after we receive the prediction.

So, in this case we perform simulations to find the optimal price to bid for a given competitor price prediction by applying samplings from the new probability function seen in Figure 6-6 against each of our possible bids. As in the case of calculating the value of perfect predictability, those given competitor price predictions still must obey the only view of the future that we *currently* have, the probability function of Figure 6-3. That is, any new means of prediction, whether perfect or imperfect, is necessarily *conditional* on our current probabilistic assessments of what we are now predicting. Therefore, the predicted prices that we use in our simulations are in accordance with the probability function of Figure 6-3.

This process is illustrated by Figure 6-7. For a given discrete slice k of the original distribution, we find the optimal bid price. While in the case of perfect predictability the optimal bid price for a given prediction was to bid just below the predicted competitor price, that is not likely to be the case here because of the residual uncertainty and its asymmetric nature, as is depicted in Figure 6-7. We move in this manner along the entirety of the original distribution, finding our optimal bid price for each prediction slice.

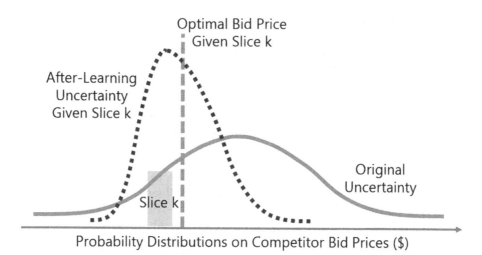

Figure 6-7. Sketch of process for determining the value of learning with imperfect predictability

Then, as we did in the case of perfect foresight, we select each possible competitor bid-price level (the prediction slices) and their probability of occurring (which is the area under the original distribution curve of each slice) and determine the associated expected profit generated by each of the associated newly determined optimal bids. Summing up over all the slices the expected profit for each slice weighted by the associated probability of each slice is the profit that is expected to be delivered through the reduction in uncertainty that is attainable by acquiring and analyzing the competitor materials costs information. In other words, the equation in the last section for determining the expected profit after learning for *perfect* predictability applies to *imperfect* predictability as well—the only difference is the way the optimal bid is determined for each prediction slice.

The difference between this expected profit of optimal bidding given the new, *reduced level* of uncertainty of our predictions about competitor bids and the expected profit of the optimal bidding given the *original* level of uncertainty about competitor bids is the *expected learning value* of partially resolving the limiting constraint on competitor bid pricing via the materials costs analysis. However, it should be noted that this expected learning value relates to just a *single* future bid situation.

Modeling Refinements

If we expect recurring bids of this type, the expected learning value will understate the full expected learning value of the competitor cost analysis, since it was calculated for only one future deal situation. To quantify the full expected learning value, we simply sum up the value of all the expected future relevant bid situations (which itself may be an uncertain variable), discounted by a time value of money factor as appropriate. And with that, we have a credible estimate of the expected value of the competitor cost analysis that will partially resolve the limiting constraint (with the caveats discussed later). We can then compare this expected learning value to the cost of acquiring the relevant data and performing the cost analysis. If (and only if!) the benefit exceeds the cost, we should make the investment—and importantly, we now have the credible quantifications to get executive buy-in to do so!

Of course, models are necessarily only approximations of the real world, and there is always the question of how much more accuracy is required when making learning-value estimates. The general rule, as always, is that if the expected marginal benefits of increased predictive accuracy are greater than the marginal costs, we should keep refining the model. Having said that, the reality is that there is usually a bit of art involved in ascertaining when "enough is enough" because of the inherent complexities involved. But we should always be guided by the "better to be approximately right than precisely wrong" motto. Better to roughly estimate than ignore.

Certain rules of thumb or heuristics can be helpful in considering modeling refinement decisions. For example, a key issue to consider in any model is whether the uncertain variable being considered is dependent on the behaviors of other people. If so, it is generally a big mistake to ignore the likelihood that these people will *adapt* to your actions, changing their behaviors and influencing the probability distributions on their behaviors that you previously used to determine your actions. For example, in our bidding example, if we alter our bidding actions based on our analysis that we have performed thus far, it is reasonable to assume that the competitor will ultimately pick up on what we are doing and alter its bidding accordingly. How long that adjustment may take is probably uncertain, and we should encode that uncertainty in a probability distribution and incorporate it in our model so that we do not grossly overestimate the expected learning value. In other words, for our example, it is almost surely wrong to simply sum up the number of expected relevant bid opportunities over a long period of time and multiply that number by the expected learning value of the very first bid opportunity for which we will apply the new pricing strategy. Rather, the expected learning value associated with each opportunity will likely tail off over time as the competitor adapts, moderating the total expected learning value that is associated with better predicting competitor materials costs.

In general, when considering uncertainties related to the behaviors of intelligent agents, we must consider that they will adapt, and we should therefore further consider if there are game-theoretic overlays that need to be included in the modeling. This issue comes up a lot, and the effect on value-of-learning calculations can be significant, so it cannot be overstated how important the distinction between adaptive- and non-adaptive–based uncertainties is when modeling uncertainties over time. Again, better to be approximately right by taking into account all the dynamics of a situation in an estimate than being precisely wrong by not doing so.

For learning that enhances the prediction of behaviors associated with *adaptive actors*, calculating the expected value of learning over multiple action instances must account for the possibility of adaptation to the actions over time.

But given that we consider the various modeling nuances that matter, the procedure outlined here will provide a good understanding of the value of addressing that most important value bottleneck of our pricing example, and this same procedure can be applied for optimizing any other data-to-learning-to-action process. While not always the case, many decisions, such as this bidding example, are explicitly described in terms of an optimization; for example, determining the optimal inventory level given uncertainties about demand and perhaps other factors. Often, optimization models are already in place in such cases that maximize the expected value *given* the current assessment of uncertainty. Those models can usually be leveraged to determine the expected value of learning for various potential solutions aimed at *reducing* the uncertainty. In other words, our value of learning approach can simply be a matter of extending existing modeling in such cases.

Having worked through an example, we will now turn to some general methods for *describing* learning and decision situations, which can be useful for more clearly thinking through complex decision situations, for identifying common learning patterns and synergies among decisions across an organization, and for communicating and collaboratively working on learning and decision models.

Learning Value Modeling

A general way to describe a decision-based situation is with a *decision diagram*, sometimes called an *influence diagram*. While they look simple, decision diagrams can encode arbitrarily complex decision situations and provide an excellent means for facilitating communications and collaborations about decision models, as well as make it easy to incorporate more details as the understanding of a data-to-learning-to-action process grows. A decision diagram is a special type of *directed graph* that includes the following elements:

- *Decision nodes* that represent decisions with respect to alternative actions and that are conventionally depicted as rectangles or squares

- *Uncertainty nodes* that represent uncertainties and are depicted as ovals or circles

- *Value nodes* that represent measures of value that are to be maximized and that are depicted as octagons

- *Directed lines terminating at a decision node* denote that the information from all the nodes that are connected by the lines to the decision node is available *prior* to the decision being made.

- *Directed lines terminating at an uncertainty node* denote that the uncertainty is *conditional* on all the nodes that are connected by the lines to the uncertainty node.

- *Directed lines terminating at a value node* denote that the value is a function of all the nodes that are connected by the lines to the value node.

Our pricing example is, of course, a decision-based situation, and it can therefore be described by a decision diagram, as is shown in Figure 6-8. The octagonal value node is the profit, if any, that will be accrued from a deal with the customer. The value node has directed lines connecting it to two other nodes of the diagram, indicating that the deal profit is a function of the two nodes. The first of those nodes is our bid price, which is a decision that we will make and is therefore represented as a rectangle. The second node that the deal profit is dependent on is the competitor bid price, which is an uncertainty and is therefore represented as a circle.

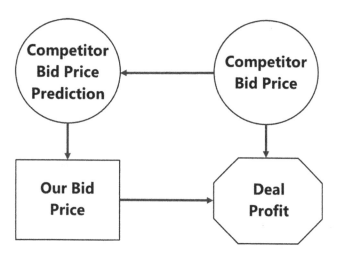

Figure 6-8. Decision diagram of the pricing example

Our bid-price decision is made with the benefit of the prediction of the competitor's bid price, as depicted by the line from the competitor bid-price prediction node to our bid-price decision node in the diagram. The bid-price prediction is an uncertainty and so is represented by a circle. Finally, the competitor bid-price prediction is conditional on the actual competitor bid price, which is, of course, an uncertainty and therefore is represented as a circle.

We will only know the competitor's actual bid price *after* we make our decision on our bid price, which is why there is no line directly from the competitor bid price to our decision node. If there *were* a direct line connecting the competitor bid price with our decision node, it would be the equivalent of having perfect foresight on the competitor's bid price, providing a value of perfect learning. But, realistically, the best we can do is to make predictions on the competitor bid and use that information when deciding on our bid price, as depicted in the diagram. Adjusting the uncertainty that we previously had with respect to the competitor's price bids based upon enhanced predictive capabilities represents learning because, by definition, updating probability distributions that embody uncertainties *is* learning. We can therefore call this predictive node the *learning variable* for this decision situation, and, as we have seen, the value of the learning associated with this learning variable can be quantified. An enhancement to traditional decision diagrams is depicted in an upcoming figure (Figure 6-10) in which these learning variables are diagrammatically distinguished from other uncertain variables.

We can easily add more detail to our decision model. For example, as shown in Figure 6-9, we can add the uncertain variable associated with competitor materials costs that will affect both the competitor's bid price and our prediction of the competitor's bid price. Competitor materials-costs prediction is another learning variable because it delivers predictions that reduce our overall uncertainty with respect to competitor materials costs, thereby facilitating better competitor bid-price predictions.

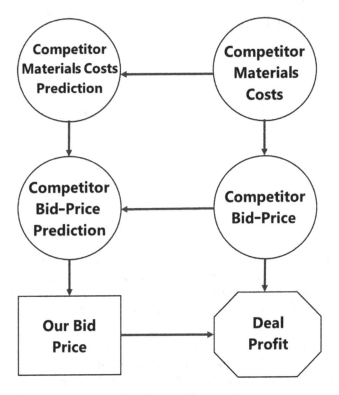

Figure 6-9. Expanded decision diagram of the pricing example

As this simple example demonstrates, there can be multiple learning variables in any given decision situation, and collectively—and in combination with the model of the decision itself—they represent learning that has a potential effect on the decision. And as we have previously discussed, determining potential solutions that reduce uncertainties associated with limiting constraints on value, potential solutions that can be represented as learning variables in decision diagrams, is an exercise that benefits from creative thinking.

We can then add a couple of other features to the diagram that go beyond standard decision diagrams and are helpful in optimizing data-to-learning-to-action processes. First, we can explicitly identify the parts of the diagram that directly represent our learning by shading the uncertainty nodes that represent the learning variables. This is useful because we particularly want to focus on opportunities for creating value through learning, and these are the nodes that represent those opportunities. Second, because we are often concerned with data-to-learning-to-action processes in which the decisions are recurring, we want to identify uncertainties that may be affected by our *prior* decisions. This will typically be the case when the uncertainty relates to the behavior of an adaptive agent, such as the competitor in our pricing example. This identification is helpful in ensuring that when we sum up the value of learning associated with updates to the learning variables, we account for this adaptation and scale the value of learning accordingly. We can simply apply an asterisk to such uncertain nodes. Figure 6-10 illustrates this modified decision diagram that we can call a *learning-value diagram*. As in the case of its decision diagram precursor, this learning-value diagram can easily be enhanced as necessary by adding additional nodes and connections.

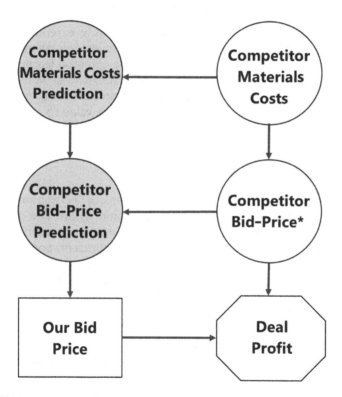

Figure 6-10. Learning-value diagram

Although not explicitly shown on the learning-value diagram, there exist models within minds and/or machines that underlay the directed lines connecting the nodes. For example, translating assessments about competitor materials costs into bid-price predictions requires a model. It could be a simple spreadsheet model or something more complex. Deciding on our optimal bid price given our less-than-perfect bid-price predictions also requires a model. These models are all as much of the learning associated with the data-to-learning-to-action process as are their informational inputs, and are therefore necessarily also a contributing part of the learning-value equation.

In fact, it is quite often the case that in working to determine an expected value of learning it is found that the *current decisions* are not optimal given the information available. The decision model that translates the available information to action may simply not be good enough. So, there might be an opportunity to apply some of the basics of modeling to the current decision to ensure that it is optimal and provides a proper baseline before even considering additional learning enhancements. As in the case of our pricing example, that could mean applying simulation techniques to determine optimal bids rather than, for instance, relying on human intuition.

As is readily seen, learning-value diagrams are easily extended as analysis progresses for a data-to-learning-to-action process, encompassing increasingly finer detail as is warranted. And where there are common learning variables, learning-value diagrams can enable an integration across what would otherwise be separate data-to-learning-to-action chains. Institutionalizing, extending, and keeping evergreen learning-value diagrams is of high value to an organization in and of itself.

Alternative Modeling Perspectives

Our learning-value diagram is a compact way to depict a decision and associated learning situation and is a derivative of a decision diagram. The other classic way to depict and model a decision is with a *decision tree*, a technique you are probably familiar with, at least in passing. It is a representation in which the decision to be made is on the left side and the tree structure expands to the right, with branching determined by uncertain variables, and with values at the ends of each branch. The *expected value* of each decision option is determined by multiplying probabilities along the branches with the associated terminal values and then summing the resulting values across the branches that correspond with each of the decision options. A sketch of a decision tree that could apply to our pricing decision is shown in Figure 6-11. Given the rough quantities in the sketch, we should bid a "Medium" price since that provides the greatest expected value.

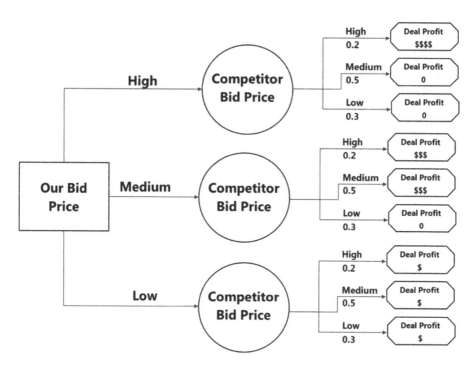

Figure 6-11. Pricing decision tree

While decision trees are always a valid decision-modeling option and can be used to calculate expected learning value, in practice they have some drawbacks. The obvious one in our case is that there is potentially a huge number of potential bids that could be made by our competitor and us, and if we tried to illustrate all the options in a decision tree, it would explode in size.

But, fundamentally, in applying a decision-tree structure, we essentially use the same technique to calculate an expected value of learning as we already discussed. We determine the action that we should take given our current learning and its corresponding expected value (i.e., the expected profit associated with a "Medium" bid price), and then we compare that to the expected value of the optimal decision given the enhanced learning. The difference is the expected value of the learning, and it will be non-zero if the additional learning serves to potentially change our decision at all.

With decision-tree formats, we place the result of additional learning (which is equivalent to the learning variables in our learning-value diagrams) to the left of our bid-price decision because we will have the benefit of this learning *prior* to making our decision. This procedure is known as *reversing* the tree. If the learning provides perfect predictability, we can easily "cherry pick" from the tree in making our decisions given a prediction. But even with a clairvoyant with perfect foresight, as we have already discussed, the prediction must obey the prior distribution (if the clairvoyant could actually change outcomes, rather than merely predict outcomes, she would be a wizard, and we would ask her to fix it so that our competitor would always bid high!), so there is a 20 percent chance the clairvoyant will predict (infallibly) that the competitor bid will be high, 50 percent chance that the predicted competitor bid will be a medium price, and a 30 percent chance the predicted competitor bid will be a low price.

As shown in the reversed decision tree of Figure 6-12, whereas before we were forced to always bid medium, now when the clairvoyant indicates the competitor will bid high, we will bid high, and when the clairvoyant indicates the competitor will bid low, we will bid low, as illustrated by the shaded outcomes in the figure. So, clearly perfect predictability in this example has value because our decision can change.

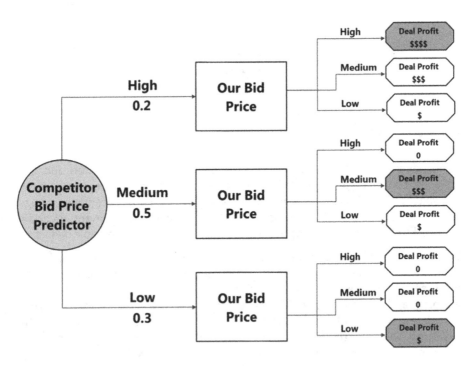

Figure 6-12. Pricing decision tree reversed and with perfect predictability

With less-than-perfect predictive accuracy, a similar procedure holds, except now when the competitor will actually bid high, which the competitor is expected to do 20 percent of the time, the prediction is sometimes wrong. For example, 80 percent of the time the predictor correctly indicates the competitor will bid high when the competitor will indeed bid high, but 15 percent of the time the predictor will erroneously predict medium, and

5 percent of the time the predictor will predict low. Similar less-than-perfect foresight occurs for the predictor's predictions of the competitor's actual medium and low bids. We can calculate out all nine pairs of these probabilities, such as the pair 20 percent multiplied by 80 percent, which is the probability that our predictor indicates that the competitor bid will be high given the competitor bid is indeed high, and 20 percent multiplied by 15 percent, which is the probability that our predictor indicates the competitor will bid medium, but the competitor bid is actually priced high.

But what we really need to know in order to calculate learning value is the probability that the predictor is going to indicate high, medium, and low. That's done by selecting the probabilities from the set of nine probabilities that correspond to the predictor indicating each of these competitor bid price levels. So, for an indication of "high," it will include the 20 percent times 80 percent equals 16 percent, but added to that are the probabilities in which the competitor price is actually medium but the predictor indicates high and those in which the competitor price is actually low but the predictor indicates high. This calculation results in the probability of receiving each of the possible predictions—all prior to making our decision. This is known as a *preposterior* probability. It represents the probability of each possible prediction *before* the predictions are delivered.

Further, we can calculate the *posterior* probabilities of the competitor's bid price that are *given* the predictor's prediction of the competitor price. Those probabilities are calculated by dividing the probabilities of each of the actual competitor bid prices occurring for each given prediction by the associated preposterior probabilities. We can then reverse the decision tree that moves the decision on our bid price to a position on the tree that is located *after* we get a prediction (to the right of the predictor's prediction). Then, we calculate the optimal bid decisions based on the posterior probabilities of each branch and the associated expected monetary values associated with each of those optimal decisions. Summing up these monetary values and subtracting the expected value of the original decision that was without the benefit of the additional, albeit imperfect, learning is the expected value of learning associated with the predictor. This is just another way of describing the same type of process that we performed using simulation previously.

The Bayesian Approach

Many readers will recognize that this procedure for determining a value of learning rests on a *Bayesian* approach, which is the general method for determining the probability of *x* given the probability of *y* when the probability of *y* given *x* is known. In our case, we knew the expected accuracy of our predictor. That is, we knew what the probability was of correctly predicting that the competitor's bid would be high (80 percent), which is the probability of predicting the competitor's bid is high *given* that the actual competitor bid *is* high. What we needed to do was to translate that probabilistic relationship into the probability of each actual competitor bid-price level given the predictor's indication. That's what the Bayesian method enables us to do.

Bayesian approaches are fundamental to learning processes in general, whether explicitly calculated in the manner of decision trees or not. Prior probability distributions that are updated to become posterior probability distributions that in turn serve as the prior distribution as new learning occurs is fundamental to our universal process, data-to-learning-to-action.

Bayesian-based techniques can be applied to handle more complex situations than our simple examples in this chapter. For example, it is quite often the case that two or more different learning variables can affect our prediction of an uncertain variable. The same techniques of encoding the decision in a value-of-learning diagram and applying a Bayesian modeling approach can handle these and other types of situations that can be arbitrarily complex.

In this book, we take pains to discuss learning and its application in data-to-learning-to-action processes for *both* minds and machines since they can complement, as well as compete for, various learning applications and activities. So, a natural question to ask is: "Does human cognition and decision making conform to a Bayesian approach?" The answer seems to be, to a limited extent, yes, but with a variety of heuristical overlays at higher levels of cognition, some of which can unfortunately serve to subvert the application of a purely Bayesian perspective.[4] Which is why structured approaches as presented in this chapter are generally necessary to augment native human learning and decision making when optimizing enterprise data-to-learning-to-action processes.

---

[4]See, for example, the following research, which is indicative of limited Bayesian processing at lower levels of cognition: Devkar, Deepna, Anthony A. Wright, and Wei Ji Ma, "Monkeys and humans take local uncertainty into account when localizing a change", *Journal of Vision* 17(Sept. 2017): 4. doi:10.1167/17.11.4

Summary

We covered quite a bit of territory in this chapter; let's review some of the key points:

- Notwithstanding our inclinations, decisions cannot be judged based on outcomes. They can only be judged based on the quality of the learning upon which the decision is made, and the learning is dependent on the state of information that exists at the time of the decision.

- Uncertainty is not an intrinsic property of an event or an object. It is an assessment based upon information and learning, whether that assessment is made directly by a human or by means of a computer-based model.

- The expected value that is associated with learning that provides *perfect* predictability of an uncertain variable, while typically unlikely to be attainable, provides a useful upper bound on the expected value of any possible solution aimed at better predicting the uncertain variable.

- The expected value that is associated with the more typical learning situation that provides *less-than-perfect* predictability about uncertain variables can also be calculated. This value can then be compared with the expected cost of the learning.

- Learning-value diagrams can encode the relevant aspects of a current decision, including the uncertainties upon which the decision depends, some of which may represent our less-than-perfect predictive capabilities. Such uncertain variables we call *learning variables* because they have the potential to be improved through enhancements of information and learning.

- It is learning variables for which we can quantify an associated value of learning. The quantification entails comparing the value of the default action given the current state of learning with the value that is expected to accrue if an existing learning variable is enhanced or a new learning variable is applied.

- If the expected learning value associated with a learning variable is greater than the cost to implement the capability represented by the learning variable, the capability is worth implementing.

- For recurring decisions of the same type, the full value of learning associated with the capability is the sum of the value of learning associated with each decision.

- Care must be taken in summing the value if the learning variable represents predictions of behaviors associated with adaptive actors.

Now that we have a process for calculating the value of learning, in the next chapter we will discuss how we can make decisions on alternative learning opportunities, as well as on mixed portfolios of learning and traditional investment opportunities.

Total Value

In the last chapter, we reviewed how the expected value of the learning that influences a decision in a data-to-learning-to-action process can be quantified. This chapter will introduce the concept of *total value*, which includes learning value, and describes how optimizing expected total value across an organization's portfolio of investment opportunities is the path to optimizing the organization's long-term performance.

Total Value

Projects and activities such as building a new manufacturing plant or hiring people are investments that are expected to generate value for the organization. That expected value is ultimately a function of the net cash flows that are expected to be generated by the investment. There is typically uncertainty associated with those future cash flows, and they must therefore be appropriately adjusted for this uncertainty, hence the term *expected* value of the cash flows. That expected value can be calculated by multiplying cash flows, or elements of the cash flows, by the probability that the cash flows or elements thereof will occur.

The present value (discounted by the cost of capital of the organization) of these expected net cash flows can be termed the expected *direct value* of a project or activity. This is the value that is calculated for potential investments using the standard economic modeling that is embedded in numerous spreadsheet models in a typical organization. These calculations of the expected value of the potential investments support decisions on whether to invest in each opportunity and can be used to rank the relative attractiveness of a portfolio of investment opportunities.

© Steven Flinn 2018
S. Flinn, *Optimizing Data-to-Learning-to-Action*,
https://doi.org/10.1007/978-1-4842-3531-7_7

These standard types of investment opportunities are distinguished from the investments that we have been discussing thus far, investments in *learning*, which are investments that are aimed at reducing uncertainties. In contrast, in standard investment modeling these uncertainties are typically taken as *givens*. These given uncertainties are (or at least, should be) reflected in probabilities that are applied when calculating the expected direct value of the project or activity. If the net present value of the cash flows, after adjusting for the probabilities, is negative, then the project or activity should not be undertaken. Learning that could potentially result in a change to those probabilities such that the decision would potentially be different than the current default of not pursuing the opportunity (i.e., the expected direct value becomes positive) of course has value, and that is the expected learning value that we discussed in the last chapter. In other words, after the learning takes place and the associated probabilities are updated, these updated probabilities are then taken as *givens* by the standard type of investment models that calculate expected direct value. In a sense, then, the expected learning value converts to expected direct value after the learning occurs.

Actionable learning is learning that has the potential to *change* the current *default* decision, and therefore has value; i.e., expected learning value. This expected learning value is then converted to expected direct value after the learning occurs.

And, again, a key point is that because of the rigorous way the value of this learning can be quantified, as was illustrated in the last chapter, *the value of investments in improved and actionable learning are just as real as those calculated for standard types of investments*. Importantly, this means the value of learning can be legitimately compared on an "apples to apples" basis with the value of other investment opportunities.

Some projects and activities have *both* an expected learning value and an expected direct value (that is *other* than just *the investments in the learning*, which we also assign to the direct value category). That is because there may be learning embedded within a project that may potentially influence subsequent decisions even if the *primary decision* with respect to the current project itself is *pre-determined*. As an example, if people with a certain university degree are hired who have different degree credentials than those of historical hires, there will be an expected direct value associated with the hire, a value that is presumably similar to, but perhaps a little less than, that of historical hires because of the risk of the hire not working out as well as the historical hiring norm. But there will also be expected learning value because the difference in credentials of this non-conventional hire will provide *new information* that may influence subsequent hiring decisions. This could be because, for example, the offered salary might be a little less for the degree held by this experimental hire, and therefore if the long-term performance of the hire is as good as that of historical hires, value will be created for the organization because subsequent hiring decisions will be influenced accordingly.

By the way, this particular example represents a common path to value creation, whether in the securities markets or in procuring assets for a business—namely, acquiring assets that are in some way undervalued by the marketplace as a whole. The identification of such undervalued assets requires learning that others, or at least most, in the marketplace do not have, and such actionable learning is, of course, valuable.

This HR-based example also exemplifies the fact that many projects and activities can include what is essentially considered embedded experiments, and those experiments can have a quantifiable learning value. In fact, these types of *embedded experiments should generally be encouraged and thoughtfully engineered into projects and activities,* and the ability to quantify the expected learning value of these prospective embedded experiments provides the means to justify those experiments that can be expected to have the greatest value leverage.

Because we have apples-to-apples value quantifications of expected direct value and expected learning value, we can not only *compare* the respective values, but we can also legitimately *combine* them. The combination of expected direct value and expected learning value of a project or activity constitutes the expected *total value* of the project or activity. The mix of expected direct value and expected learning value for a given project or activity can vary from extremes in which the expected learning value is negligible to the opposite, in which expected learning value comprises nearly the entirety of the expected total value. Expected total value therefore represents the true, full value of a project or activity and is therefore what should be used to make investment decisions on all of an organization's prospective projects and activities. This is a significant change—and one for the better—from most current practices!

Expected Total Value = Expected Direct Value + Expected Learning Value

---

Business-Renewal Lifecycle

The relative mix of expected direct value and expected learning value of projects and activities tends to vary with the projects' and activities' positions within an organization's *renewal lifecycle. Business renewal* is a term that refers to all the projects, processes, and activities in an organization that are associated with developing new ways of working, new business approaches, new markets, and new products and services, as well as the retirement of older approaches, markets, products, and so on. In other words, it is all the projects and activities associated with the continuous adaptation to current and anticipated circumstances. In short, *renewing* the business, just as the cells in your body are continuously renewed. It's everything other than just business-as-usual activities, and thus effective business renewal is key to sustainably excellent business performance. As you will have no doubt already anticipated, the punchline is that business renewal is all about *learning*.

Business renewal–related projects and activities in an organization follow a distinct *lifecycle*, as illustrated by Figure 7-1. The horizontal axis of the figure represents expected direct value, and the vertical axis represents expected learning value. In both cases, these values are in terms of the present value of expected cash flows. While expected direct value can be negative, expected learning value can only be zero or positive. That's because information and the associated learning cannot destroy value (assuming the learning is validly performed) and because we are allocating investments of all types, including those that generate learning, either directly or as a by-product, to the expected direct value category.

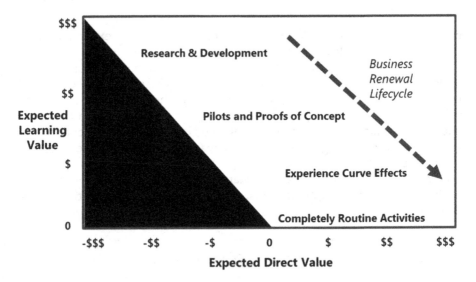

Figure 7-1. Business-renewal lifecyle

The shaded portion of the diagram represents projects and activities that have negative expected total value; that is, the associated expected learning value is insufficient to overcome an expected negative direct value (which could include the expected costs of the learning). Of course, these are projects and activities that shouldn't be undertaken. Or, if they are in progress because they were originally expected to have positive total value, but now looking forward, the expected total value is negative, they should be shut down (sunk costs are sunk costs!).

The business-renewal lifecycle begins with projects and activities that are primarily about learning, and so expected learning value is the dominant value. We discussed earlier the label we generally apply to those activities in which *all* the value is learning value: *experiments*. And the general business term for functions that comprise activities that are dominated by the performing of experiments is *research and development*, which is positioned at the top left of Figure 7-1. We often automatically think of R&D as being oriented toward new products, but R&D can apply to any functional or process area. For example, research into the concept of hiring employees outside of the typical characteristic and credential profile, as previously discussed, would be an example of HR-based R&D. Typically, there is no expectation of value generation other than learning value for what are fundamentally R&D activities—the "expected direct value" is simply the investment required to fund the R&D, therefore the expected direct value is negative. So, the expected total value is just the expected net value of learning at this initial stage of the renewal lifecycle. But this expected total value still must be positive!

The trajectory of the business-renewal lifecycle is from the top left of Figure 7-1 to the lower right, with expected learning value naturally becoming an increasingly smaller portion of the total value as it converts to positive expected direct value as learning is completed and the lifecycle progresses. Moving along the business-renewal lifecycle past R&D are pilots, demonstrations, and proofs of concept (which I'll just refer to as "pilots"). These types of initiatives are often based on results of earlier R&D activities in which the associated learning reduced uncertainty about a potential new approach, but left some residual uncertainty that needed to be addressed before committing fully to implementing the new approach. Pilots are ways in which that residual uncertainty can potentially be sufficiently reduced to enable a decision to go forward on a more significant implementation of the new approach, or, alternatively, a pilot may reveal unforeseen issues that would destroy value if a more significant implementation of the new approach were to proceed.

In some cases, pilots may have a positive expected direct value in addition to an expected learning value. In other words, the pilot, particularly if it is assumed *a priori* that it is likely to be successful, may have a positive expected direct value. In such a case, the valuable expected learning of the pilot would be in revealing any potential "show stopper" that would change the default decision of going ahead with a scaled-up implementation. By way of contrast, for a case in which the pilot is, say, a scale model of the ultimate production process that would be destroyed after testing is completed, the pilot would have negative direct value.

Our HR example could include a pilot phase. Assume that the organization has traditionally always hired graduates of computer science programs, but that the initial R&D that was conducted, consisting of surveys and analysis of other organizations' hiring practices, suggested that certain humanities majors, particularly philosophy majors, compared favorably to computer science majors with respect to long-term job performance. And research further indicated that candidates with such majors could be acquired at less cost because they were currently less in demand for software-related jobs. Hiring a few philosophy majors would be an example of a pilot that would have a positive expected direct value in addition to an expected learning value, assuming that the hired employees were expected to at least deliver a level of value that exceeded the associated salary and benefits.

The relationship between R&D and pilots for this HR example, in which uncertainty is reduced by an initial, relatively inexpensive action (e.g., surveys and analysis) but significant residual uncertainty remains, which then prompts attempting to further reduce uncertainty by applying other, often more expensive actions, is a common situation. Given the residual uncertainty after the initial R&D, rather than a full commitment to, say, the hiring of dozens of philosophy majors, another action, a pilot, is conducted that is expected to further reduce uncertainty, but with less value risk than jumping right to the full commitment.

Such sequencing of step-wise reductions in uncertainty constitutes an example of *experimental design*, whether it is formally considered that or not, and the overall renewal lifecycle can be thought of as following an experimental-design trajectory. As we move along the renewal lifecycle, there is a natural *sequence* of decisions in which uncertainty is reduced but not eliminated early in the lifecycle, and then reduced further, but not completely eliminated, at each subsequent phase of the lifecycle. For example, the learning occurring at the R&D stage enables more-informed decisions on whether to conduct pilots, and the learning occurring at the pilot stage enables more-informed decisions on whether to go forward with full-scale implementations.

Essentially what is happening at each stage of the renewal cycle is that learning is being performed that serves to reduce the equivalent of false positives and false negatives with respect to subsequent stages, or what you may remember from statistics classes as type I and type II errors, as is illustrated by Figure 7-2.

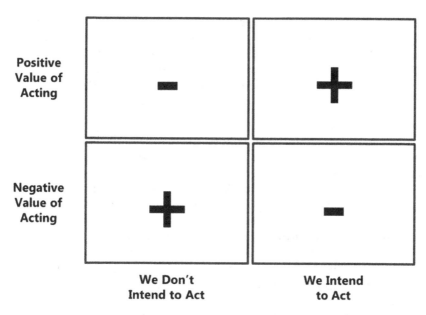

Figure 7-2. Action intention matrix

The business-renewal equivalent of a false positive or type I error is the situation in which we *intend* to take an action (i.e., taking the action is the *default* decision) at the next stage of the lifecycle but we *shouldn't*, and the action therefore has negative value (bottom-right quadrant of Figure 7-2). And the business-renewal equivalent of a false negative or type II error is the situation in which we *don't intend* to take an action (i.e., the default decision is *not* taking an action) at the next stage of the lifecycle but we *should*, which results in foregone, and therefore negative, value (upper-left quadrant of Figure 7-2). While we don't usually think of renewal-based processes and activities in these experimental design–related terms, it is really exactly the same idea, just writ much larger in the case of the overall renewal lifecycle.

Proceeding along the renewal lifecycle, the renewal stage after the pilot stage is composed of projects, processes, and activities for which there are *experience-curve* effects. These are fully production activities, but for which actionable learning continues. The expected direct value dominates the total value of these activities, but there is also some expected learning value as well. These constitute the bulk of the projects and activities of most businesses.

As the term *experience curve* suggests, the learning value for these areas tends to result in *improvements* that are based on experience rather than completely new directions. In the classic notion of the experience curve, efficiency increases with the cumulative volume of products produced.[1] Of course, more generally, the experience-curve concept translates to services and processes in general, as well as product production.

As with other management-related concepts, our optimizing data-to-learning-to-action approach and more robust perspective on learning provides foundational insights on the classic experience-curve concept. Improvements based on experience are clearly a result of ongoing learning that influences sub-decisions within the overall activity. As an example, after passing the R&D and pilot stages of the renewal lifecycle it may be decided to go ahead and bring a new product onto the market. However, the product has numerous individual parts that are sourced from different suppliers. Even after the production and sale of many of these products, there will be myriad ongoing decisions associated with the specific parts that are to be used and the vendors that are chosen to supply the parts, and there will be value associated with the actionable learning that can improve those sub-decisions. There will also no doubt be ongoing sub-decisions on specific manufacturing techniques, such as automation options, marketing approaches, and so on, and actionable learning that could potentially change these decisions will have tangible value. The positive economic effects of the experience curve are really just the effects of actionable learning from the experience.

Given that these experience effect–based types of processes and activities make up most of the processes and activities of most organizations, there is a significant learning-value upside. Unfortunately, historically that upside has often not been fully recognized. For activities within formally designated R&D organizations, it is quite natural to be thinking about what more can be beneficially learned. But in more mature, production areas of a business, the "autopilot" can be set, and opportunities for learning are therefore missed. There is almost always value in conducting mini-R&D and pilot activities even in mature processes.

Again, our HR scenario is an example of such an embedded pilot with the aim of gaining actionable learning. Hiring is a mature process for most organizations. In our example, computer science majors have *always* been hired. Probably that has been the case for years. But running a mini-pilot within this mature process by hiring a philosophy major or two could yield insights from the associated learning that have the potential to deliver significant value.

---

[1]Ghemawat, P., "Building Strategy on the Experience Curve", *Harvard Business Review*, March–April 1985.

Explicitly engineering opportunities for learning even in mature processes is an important opportunity for business value creation.

In addition to the autopilot mentality, the other reason that explicit learning opportunities are typically under-performed outside of areas that are more explicitly recognized as R&D is because of the brutal reality that if the expected value of an activity cannot be credibly quantified, it will likely struggle to gain approval. Or, if initially approved, it may struggle to stay alive. Of course, this book is meant to help solve that problem!

At the end of the renewal lifecycle depicted by Figure 7-1 are completely routine processes and activities. These are processes for which there is little to no uncertainty that would affect the outcomes of sub-decisions. These types of decisions can essentially be run in an automated fashion. If there are people involved in such rote processes, the opportunity or the threat, depending on your perspective, presented by the popular meme "the robots are coming" is an inevitability.

However, it is rare for *any* process to be completely immunized from uncertainties, so we must not be too hasty to relegate a process to this no-learning zone. For example, while perhaps the current *operational* decisions of a routine process may be perceived to gain little value from additional learning, there may be ongoing decisions on the *directions* with respect to the overall process itself. In other words, even if the "how" of a mature process is completely routine, there may be decisions and valuable ongoing learning to be had with respect to the "what" and the "why" of the process.

Recall from the last chapter that we can use learning-value diagrams to model the uncertainties and associated learning variables that are related to a decision. Figure 7-3 depicts different positions of the renewal lifecycle in terms of conceptual learning-value diagrams. Earlier in the lifecycle, the learning-value diagrams will have more learning variables (the shaded circles of Figure 7-3) associated with uncertainties (the non-shaded circles). Later in the lifecycle, there will be fewer uncertainties and learning variables, but only rarely will the uncertainties and learning variables completely vanish.

Earlier in Renewal Lifecycle   **Later in Renewal Lifecycle**

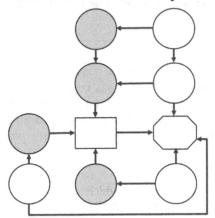 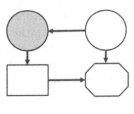

Figure 7-3. Contrasting conceptual learning-value diagrams at different points in the business-renewal lifecycle

Regardless of the position in the renewal lifecycle, the question of whether to take action to reduce the outstanding uncertainties remains the same: "Does the expected value of learning that can potentially reduce uncertainty and thereby affect the current default decision exceed the costs of the learning?" Or, said another way, "Is the net present *expected total value* of the action positive?"

Optimizing Total Value

An organization will have many different investment opportunities that span across the renewal lifecycle. The value of some of these opportunities will primarily be attributable to actionable learning, while in some cases it will be almost all expected direct value, and there will be cases in between. Fortunately, the quantification of expected learning value as a component of expected total value enables expected total value to be the *consistent arbiter* of value for any potential investment opportunity in the organization. If the net present expected total value of an investment opportunity is positive, then the opportunity should be pursued. All the real work is in credibly quantifying the expected learning value and the expected direct value of the opportunity. Once that is done, the go/no-go decision should be very straightforward.

If the expected total value of an investment opportunity is positive, the opportunity should be pursued.

However, it must be recognized that given the reality of budgeting considerations, the decision is not always quite that straightforward. That is, budget constraints may force an organization to *triage* opportunities to fit within a budget that have positive expected total value and which all would therefore be expected to create shareholder value if pursued. It's a shame, but budgeting isn't going away, so we need to at least make the best of it by prioritizing based on *expected total value*.

An approximate approach for performing such an opportunity prioritization is to apply an equivalent to the well-known concept of return on investment (ROI), in which the expected value of the opportunity is divided by the investment that is required to attain the value. The opportunities are then ranked by ROI and cumulatively selected until the budget is exhausted. In our case, we simply use expected total value, and thus have an expected total value return on investment (TROI) we can use as the means for opportunity prioritization and selection.

But that approach only approximates the optimizing of the mix of opportunities to fit within the budget constraints and is particularly problematic for the realistic situation in which investments in opportunities and the associated investment budgetary limits span multiple years. More precision can be gained for real-world investment and budgeting situations by establishing a simple opportunity portfolio optimization model, which can be easily set up and performed with a basic spreadsheet-based optimizer.

Figure 7-4 conceptually depicts a portfolio of investment opportunities for a series of projects. The expected direct values and expected learning values of each of the projects are shown, as is their sum, the expected total value. Each of the projects has an investment requirement profile over a three-year period, and there are investment budget limits for each of the three years.

Project Name	Expected Direct Value	Expected Learning Value	Expected Total Value	Year 1 Investment	Year 2 Investment	Year 3 Investment
Project A	-10	20	10	8	2	0
Project B	40	0	40	10	10	10
Project C	-35	60	25	10	20	5
Project D	20	5	25	15	5	0
Project E	-60	100	40	20	25	15
Project F	5	5	10	3	3	0
Project G	-20	25	5	5	5	10
Project H	15	0	15	10	0	0
Project I	-5	25	20	0	4	1
Project J	-90	150	60	40	30	20
Project K	100	10	110	20	30	50
		Annual Budget Limits		90	90	100

Figure 7-4. Example project portfolio and budget limits

Figure 7-5 shows the individual project investment decisions after the portfolio of opportunities has been optimized to maximize the expected total value of the opportunity portfolio while obeying the annual budget limits, as well as shows the associated budget residuals for each year.

Project Name	Investment Decision	Optimized Total Value	Optimized Y1 Investment	Optimized Y2 Investment	Optimized Y3 Investment
Project A	YES	10	8	2	0
Project B	YES	40	10	10	10
Project C	NO	0	0	0	0
Project D	YES	25	15	5	0
Project E	YES	40	20	25	15
Project F	YES	10	3	3	0
Project G	NO	0	0	0	0
Project H	YES	15	10	0	0
Project I	YES	20	0	4	1
Project J	NO	0	0	0	0
Project K	YES	110	20	30	50
Total		**270**	**86**	**79**	**76**
Budget Residual			4	11	24

Figure 7-5. Optimal project decisions given the budget constraints

Variations of this type of modeling approach ensure the best possible allocation of capital in an organization given the inherent economic distortions that budgeting inevitably causes. But failing to quantify the value of learning would cause even greater economic distortions. Better to be approximately right than precisely wrong, and ignoring learning value in investment portfolio optimizations, as is currently too often the case, is an example of being precisely wrong!

Retrospective Value

I have generally been careful, perhaps to the point of awkwardness, to use the term "expected" with respect to learning value, direct value, and total value to ensure that it is understood that these are *prospective* concepts. So, for expected learning value, the value is what we expect to attain *if* we perform the learning. But what if we want to determine the value of learning *retrospectively*, after the learning has already been performed and the *results* of the learning have become available?

Calculating retrospective learning value can be a valuable exercise that helps you to understand the value that learning *has delivered* and as a means to effectively convey that understanding of the value to others. This quantification of learning retrospectively can be performed by again taking heed of that maxim of decision analysis that learning only has value *if it changes a decision*. In the retrospective case, we no longer have *potential* changes to decisions, as in the case of expected learning value. Retrospectively, either the learning *was actionable or not*; i.e., either it led to a change in actions that otherwise would have occurred without the benefit of the learning, or it did not.

So, to determine a value for learning retrospectively we first determine what actions have changed, or certainly will change, because of the learning that has already occurred. Then, we compare the value of these new actions to the value of the actions that would have otherwise occurred if the learning had not happened. Typically, this quantification of the retrospective value of learning is a simpler process than quantifying *expected* learning value because we don't need to worry about probability distributions and the like. But it does require an *honest assessment* of what decisions really changed based upon the learning, and this is more assured if the prospective modeling of the learning value was previously performed. And, of course, we should not simply extrapolate the *continuing* value of the subject learning from this retrospective learning value rather than performing a proper prospective analysis.

However, a major caution on calculating the value of learning retrospectively is to be mindful of that other maxim of decision analysis that *the correctness of a decision cannot be evaluated based on its outcome*. Good decisions can lead to bad results, and vice versa. Uncertainty happens. On average, though, over enough individual learning opportunities, we should get results that are pretty aligned with our original expectations. If not, we have some learning to do—we will need to examine if our expectations were at fault or if we have simply been a victim of very bad luck (of course, luck can go the other way too, but fewer questions are usually asked when that happens!).

The *expected* value of learning has not traditionally been quantified, and quantifying this value is a core focus of this book since, as I have argued, failure to quantify this value is a root cause of sub-optimal business performance. However, the *retrospective* value of learning has not fared much better in most organizations, at least in any rigorous form, and so it has been rare for organizations to appreciate in financial terms the value learning has delivered. Unfortunately, a fundamental law of management is: that which is not understood by management is inevitably made smaller so that there is less to not understand. So, learning is chronically underappreciated and underfunded in many organizations. Quantifying learning both prospectively and retrospectively is the way out of this management trap.

Summary

To summarize this chapter, we discussed the concept of expected total value, which is composed of expected direct value and expected learning value. The expected learning value can range from the entirety of the expected total value, in the case of experiments and R&D, to a negligible portion of the total value for very routine projects and activities. We reviewed the concept of the business-renewal lifecycle, in which projects, processes, and activities naturally follow a trajectory from learning value composing the bulk of the total value to learning value becoming a much smaller proportion of the overall value. Investment opportunities in which expected total value is positive should be pursued. If there are budget constraints so that opportunities with positive expected total value must be prioritized, an optimization model can be applied that maximizes the total value of the portfolio of opportunities subject to the budget constraints. And we discussed that learning value can be calculated both prospectively and retrospectively, and that both quantifications play important roles in ensuring optimal investments in learning are made and sustained.

Optimizing Learning Throughput

We have now covered the basic concepts required to optimize a data-to-learning-to-action process, but we still have a few things to consider before our method is complete. Let's recap what we have already covered before we turn to those finishing touches:

- We have reviewed the data-to-learning-to-action process and its major elements as they apply to organizations: Data Acquisition, Data Filtering, Information Management, Search and Discovery, Predictive Analytics, Process and Collaborate, and Decide and Act. We also touched on where some of the common enterprise technologies map to the chain of elements.

- We discussed how to work backward from value drivers to the associated primary decisions to determine the opportunities for enhancing learning that have the highest leverage on value. We then covered working backward from each of the primary decisions along the data-to-learning-to-action chain to identify the limiting constraints on learning that, if resolved, would deliver value.

© Steven Flinn 2018
S. Flinn, *Optimizing Data-to-Learning-to-Action*,
https://doi.org/10.1007/978-1-4842-3531-7_8

- We reviewed how to quantify in financial terms the expected value of fully or partially resolving learning constraints in a data-to-learning-to-action process, and how to generally model learning-enhancement opportunities using learning-value diagrams.

- We also emphasized the fact that because our quantification of expected learning value is every bit as rigorous as that performed for any other investment opportunity, this expected learning value is directly comparative, as well as additive, to any other calculated value for such opportunities. Expected total value, which can be partially or fully composed of expected learning value for any given investment opportunity, is therefore the proper metric to optimize across an organization's portfolio of investment opportunities.

With these concepts in place, we have the basic tools to enable optimizing data-to-learning-to-action processes and to prioritize across these optimization opportunities, as well as to effectively prioritize with respect to all other investment opportunities.

Additional Strategies

In practice, however, we need to be able to handle some additional aspects of optimizing data-to-learning-to-action that we either have not yet covered or have only mentioned in passing, as follows:

- Determining when to work on learning constraints that are *within* a data-to-learning-to-action chain, or that span *across* data-to-learning-to-action chains, *in parallel* versus sequentially

- Identifying *learning synergy* across multiple data-to-learning-to-action chains and accounting for the resulting value if it is a significant factor

- Looking ahead to the *next* limiting constraint(s) when considering solutions for the *current* limiting constraint

- *Anticipating capabilities* that are beyond those currently available when considering solutions to learning constraints

- Identifying situations for which, rather than debottlenecking a learning constraint, it is appropriate that *excess capacity* in other parts of the data-to-learning-to-action chain be *rationalized*

These additional strategies are particularly focused on optimizing learning *throughput*—and, more specifically, *actionable* learning throughput.

Parallel Versus Sequential Debottlenecking

The question of whether to address learning constraints in parallel versus just sequentially arises both *within* a specific data-to-learning-to-action process as well as *across* data-to-learning-to-action processes.

Let's start with an example of a parallel debottlenecking *within* a data-to-learning-to-action process, and then we'll generalize from that situation. As we have emphasized, within a single learning flow along a data-to-learning-to-action chain it ordinarily makes sense to only focus on working to resolve the *limiting* constraint since adding capacity anywhere else in the learning flow will deliver no value—the added capacity will be useless without some degree of resolution to the limiting constraint.

But we have discussed a case in earlier chapters (starting in Chapter 5) that exemplifies the situation in which it can make sense to address learning constraints *in parallel* within a data-to-learning-to-action chain. In that case, we worked backward along a data-to-learning-to-action chain from a pricing decision. We found at the Process and Collaborate stage of the data-to-learning-to-action chain that some deals had been lost that otherwise would have been expected to be won, and that this had happened because of the amount of time required to come to a consensus on a decision. That was the result of the awkwardness of the back-and-forth via emails as the parties to the decision tried to reach a consensus decision. This learning constraint, in this case a constraint within the process of *collective learning* among parties to the decision, can be embodied in a probabilistic form as the probability that a bid that *should be* submitted is, in fact, *not* submitted.

We found that this probability of decision failure due to a breakdown in the group decision-making process is not related to the input information that the parties consider in trying to make their decision. That is, the probability of failing to submit a bid on time is the same regardless of the inputs from upstream elements, and hence the probability of not effectively submitting a bid is *independent* of the specific nature of the inputs. This limiting constraint is therefore simply a function of the internal operations of the Process and Collaborate element itself, affected perhaps by factors such as the technology being used (e.g., standard email), intensity of travel demands on the participants, and so forth.

It would clearly be valuable to reduce this probability, which amounts to at least partially resolving this value bottleneck. That was the *first* limiting constraint on learning that we found for this data-to-learning-to-action process, and it was specific to the Process and Collaborate element (with some overlap with Decide and Act). But we also found that deals were lost because of the current limited understanding of how the competitor would bid. When we worked back along the data-to-learning-to-action chain, we found that a lack of understanding of the costs of materials within the competitor's solution was the limiting learning constraint that needed to be addressed to decrease the uncertainty of how competitors would bid. So, that was the *second* learning constraint of the data-to-learning-to-action process related to bid pricing, and it was *independent* of the first constraint on its effect on our bidding decisions. This specific situation is illustrated by Figure 8-1, which conceptually displays learning flow throughput levels along the data-to-learning-to-action chain as represented by the degree of separation between the pairs of lines that represent each individual learning flow, including the throughput pinch points (a narrowing of the separation between the lines). In our specific price-bidding example, there are two parallel learning flows, and the independent limiting constraint in each of the flows is indicated by the dashed circles.

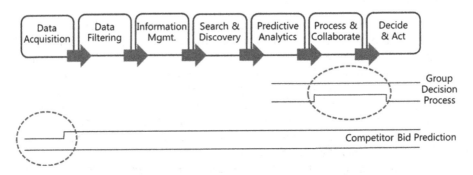

Figure 8-1. Independent learning constraints within a data-to-learning-to-action process

So, which of the constraints should we work on? If there exist solutions that can relax *each* of the constraints such that the benefits exceed the costs (i.e., expected total value is positive) and the required resources to do so are available, *both* learning debottlenecking opportunities should be pursued, subject to the caveat in the next paragraph. Otherwise, if investment resources are constrained, then the opportunities can be compared within the context of an optimization of the expected total value across a portfolio of opportunities, as discussed in Chapter 7.

A subtlety to note here is that even if learning flows and their associated limiting constraints are considered to be independent, as was the case in this example, the *value* generated by resolving the constraints may have a *dependency* and should be taken into consideration. That this can be the case is readily seen by performing a thought experiment with our example here. Specifically, assume that the resolution of the competitor bid prediction–related limiting constraint resulted in perfect predictability of the competitor bid. As we discussed in Chapter 6, we would then always profit on every bid by simply bidding a tiny bit less than the predicted competitor bid. That would mean that resolving the group decision process would then be *even more* valuable because every bid that was not made would for certain constitute a foregone profit, whereas originally in some bid instances it wouldn't have mattered because we would have lost the bid anyway. In other words, the probability that a bid that was not effectively submitted would *otherwise* win the deal and hence would generate value if it were submitted is now a certainty, whereas before it was not. Therefore, it can be useful to check the expected total value of resolving each independent constraint within a data-to-learning-to-action chain *given* the expectation that other independent constraints are resolved.

Generalizing for learning constraints that are *within* a data-to-learning-to-action process and are associated with learning flows that are independent of one another (or close enough to being independent for practical purposes), if resources permit there is generally no reason not to pursue the debottlenecking of the learning constraints in parallel if there are solutions for each that have positive expected total value. However, within a single data-to-learning-to-action chain, there may still be value dependencies even for independent learning flows that should also be considered.

The question of whether to debottleneck constraints *across* data-to-learning-to-action processes is also basically a question of resource availability and the competition for the resources, and is even simpler in concept. From a practical standpoint, resources are always ultimately constrained to at least some degree. That's where the value-driver approach comes in, as it essentially triages opportunities, facilitating an identification of, and then a subsequent deeper dive into, the highest-leverage data-to-learning-to-action processes and associated constraints.

So, in general, if the necessary resources are available, it makes sense to work to optimize multiple data-to-learning-to-action processes in parallel. And if resources become constrained, making priority decisions based on expected total value is always proper. It should also be remembered that just because a learning constraint is identified, it doesn't mean that there is a cost-effective means to address the constraint, at least in the near term. So, an organization may have a number of identified constraints that have to be put on the "back burner" in favor of addressing other learning constraints that *can* be cost-effectively addressed in the near term. It is important to keep monitoring the items on those back burners as technology advances; we will discuss this more later in this chapter.

Learning Synergies

It can be the case that reducing an uncertainty can generate learning value for *more than one* data-to-learning-to-action process. In such cases, a learning de-constraining solution that reduces the uncertainty should accrue the value-add that it contributes to *all* the relevant data-to-learning-to-action processes for which the reduced uncertainty is relevant. Because of this potential for *amplified value* of de-constraining learning, it can be beneficial to perform at least a first-cut modeling of learning flows in multiple data-to-learning-to-action process so that learning synergies across the data-to-learning-to-action processes are not missed.

Figure 8-2 illustrates a common learning constraint that applies to two separate data-to-learning-to-action processes, A and B, with the constraint occurring at the Data Acquisition element of both processes. The value of resolving this constraint is the value of resolving the constraint for data-to-learning-to-action process A *plus* the value of resolving the constraint for data-learning-to-action process B. Of course, these values may not be the same—the reduction in uncertainty could have higher value leverage in one data-to-learning-to-action process than in the other. Additionally, it can be the case that neither of these values by themselves would be sufficient for the benefits to outweigh the costs of reducing the uncertainty, but together they do (i.e., the sum of the expected total value across chains is positive). That economic reality is why it is so important to identify learning synergies—and to do so as early as possible.

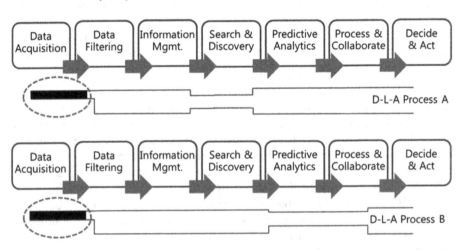

Figure 8-2. Common learning constraint across multiple data-to-learning-to-action processes

The identification of learning synergies across data-to-learning-to-action processes is alternatively illustrated in the form of a learning-value diagram in Figure 8-3. The synergies are determined by identifying common learning variables (designated in this case by black circles) across otherwise distinct learning-value diagrams. This represents another advantage of consistently using learning-value diagrams to model data-to-learning-to-action processes—it makes it easy to identify learning synergies across chains.

Data-to-Learning-to-Action Process A **Data-to-Learning-to-Action Process B**

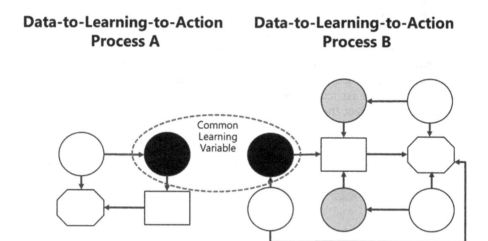

Figure 8-3. Common learning variable across data-to-learning-to-action processes

As we have discussed in previous chapters, many technologies, as well as human resource–based capabilities such as skill sets, will apply to multiple data-to-learning-to-action processes in an organization. Without adequately understanding learning-synergy value, optimal investment in such technologies and human resources is unlikely to occur. So, again, opportunities for learning synergies should be sought early in the learning-improvement process. For example, as high-level learning-value diagrams are being developed, common learning variables should be identified.

Constraint Look-ahead

While the focus on learning constraints must necessarily be on the *limiting* constraints since increasing the actionable learning capacity anywhere else in the data-to-learning-to-action process will not increase actionable learning throughput, there are often advantages to *looking ahead* to the upcoming *sequence* of potential bottlenecks in the process after the limiting constraint is resolved. This can be the case, for example, because it takes resolving more than one constraint in a data-to-learning-to-action process to generate positive expected total value. It can also be due to learning-capacity granularity or "chunking" considerations.

For example, in Figure 8-4 we have four distinct learning capacities along a data-to-learning-to-action chain, with the portion of the chain around the Data Acquisition element (labeled *1*) being the limiting constraint. Ordinarily, we would look at resolving that constraint to a level such that it meets the capacity level of the next limiting constraint, in this case labeled *2*, which is located at the Data Filtering element. However, it may be that there is insufficient value to proceed with resolving constraint *1* if the overall process is still limited to capacity level *2*. But, if we can get to, for example, capacity levels *3* or *4* there may be positive expected total value. In that case, we need to compare the expected total value (i.e., expected learning value of increased learning capacity net of the associated costs) of increasing capacity to level *3* and to level *4*, and then select the option with the highest expected total value.

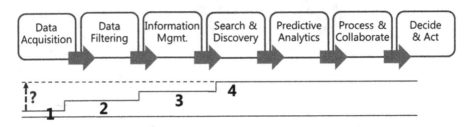

Figure 8-4. Learning-constraint look-ahead

Alternatively, learning-capacity increments may come in discrete quantities or chunks that exceed merely moving from learning-capacity level *1* to level *2*. Further, it may be the case that because of this chunking, it is not cost effective to upgrade the capacity of the Data Acquisition element if constraint level *2* is maintained at the Data Filtering element, or perhaps even if constraint level *3* at the Information Management element is maintained. So, we need to examine the costs of resolving constraint *2*, and maybe even *3*, and determine if that associated *overall* upgrade of the data-to-learning-to-action process has a positive expected total value. Of course, it could be that there are chunking considerations with respect to upgrading capacity in these other areas of the chain as well. So, clearly it can get complex.

Fortunately, we can handle any arbitrary level of constraint look-ahead complexity, even with an additional layer of complexity related to learning synergies, with an expected total value optimization approach of a similar nature to that which was illustrated in Chapter 7. In any such cases, we are essentially wanting to find the optimal "packing" of options that fit with our budget and/or resource constraints. The real effort is in developing the expected total values for each option—once that is done the optimization part is straightforward. No need for short cuts on the optimization step—put those compute cycles to use!

Anticipated Capabilities

In classic applications of resolving constraints of flows such as in manufacturing settings, once the true limiting constraint is identified (which, again, can be quite non-intuitive), determining a solution to relax the constraint is typically straightforward (although it may not necessarily be sufficiently cost-effective to implement); for example, adding another machine at the limiting constraint to operate in parallel with the similar, existing machine.

However, for our more general case of resolving constraints on actionable learning, and particularly in view of the rapid advances in technology, we need to always be mindful of the spectrum of new capabilities that can be expected to become available over time (the enterprise architect role can be an important contributor in this regard). It's why it is important to continue to monitor opportunities to address learning constraints that we "put on the back burner" as we discussed earlier in the chapter. And it's not just technology for which there might be a significant shift with respect to the comparative capabilities that can be expected to be available in the future versus what is available to apply to learning constraints now; for example, a type of skill that is required to increase learning capacity might only become available sometime in the future.

Most generally, the issue of anticipated capabilities when addressing data-to-learning-to-action bottlenecks can be encapsulated by the question of whether it is better to *wait* for improved and/or lower cost capabilities than to act *now* with current capabilities, all the while also considering the dependencies with the other elements of the data-to-learning-to-action chain. In other words, it's really another layer of consideration with respect to the basic look-ahead issue that we just discussed in the last section.

The answer to how to handle anticipated capabilities generally hinges on the *granularity* of the capabilities and their *compatibilities*. In other words, do the anticipated capabilities *build on* what is already implemented? Or do the new capabilities essentially *replace* the near-term capabilities? And if they will serve to obsolete current capabilities, what is the expected timing of feasibly implementing the anticipated capabilities? If there is a graceful compatibility with current capabilities, the methods we have already covered can easily handle that situation because it simply makes sense to go ahead with capabilities that have a positive expected total value in the near term and then add capabilities that will deliver additional expected total value as soon as they are available (assuming there are no constraints elsewhere in the chain that would limit the value delivered by the added capabilities).

However, in situations in which new capabilities would *obsolete* the near-term capabilities, we need to consider whether it is better to wait for the new capabilities rather than immediately investing in a current, more limited solution to address a learning constraint, as illustrated in Figure 8-5.

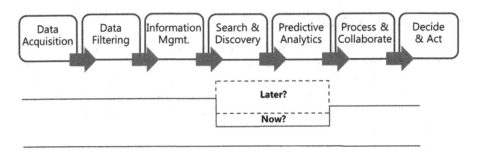

Figure 8-5. Choice of implementing a learning debottlencking solution now or a more robust one later

Determining the answer to that question is once again a matter of modeling both scenarios and choosing the option with the highest expected total value. There may be chunking issues and synergy effects to consider with each scenario, which means that there may be many sub-scenarios that need to be considered. For example, in the scenarios illustrated by Figure 8-5, the future solution, as designated by the capacity increase that is labeled *Later?*, provides more learning capacity than can be effectively used by the rest of the data-to-learning-to-action process given the current constraints throughout the chain, and so sub-scenarios should be evaluated that include increasing the capacity of the other elements of the chain that would enable the future solution under consideration to add value beyond that which would accrue by just resolving the limiting constraint.

For example, as depicted in Figure 8-6, there might be a sub-scenario, designated by *1*, in which, given that the future solution (designated by *Later?*) is applied to the current limiting constraint, the next limiting constraint (situated at the Process and Collaborate element) is brought up to the level of the third limiting constraint (which is the current learning-capacity level of the upstream elements Data Acquisition, Filtering, and Information Management). There could also be a sub-scenario, designated by *2*, in which all the rest of the elements of the chain are de-constrained up to the level of the future solution for the current limiting constraint. And, of course, there could be sub-scenarios that fall in-between or outside these two sub-scenarios.

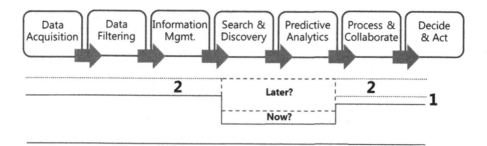

Figure 8-6. De-constraining scenarios in anticipation of future solutions

Regardless of the number of sub-scenarios that we need to examine, each sub-scenario can simply constitute a discrete option that is evaluated by our optimization model, and the option with the greatest expected total value is the winner. Again, any level of complexity can be decomposed and effectively dealt with in this manner.

Rationalizing to the Limiting Constraint

Everything we have discussed so far about learning constraints has been underpinned by the tacit assumption that we want to find ways to *overcome* the limiting factor and thereby *increase* the throughput of actionable learning in a data-to-learning-to-action process. And, indeed, this assumption is valid for most real-world situations. But there can be cases in which it makes sense to take the limiting constraint as a *given*, and then *reduce* costs by rationalizing extra learning-capacity level elsewhere in the data-to-learning-to-action chain that is above the limiting constraint's capacity, as is depicted in Figure 8-7.

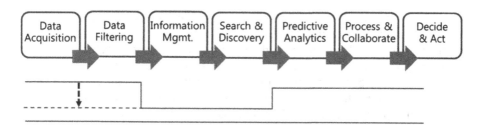

Figure 8-7. Rationalizing learning capacity to the limiting constraint

This situation can arise, for example, when it is determined that there is simply not sufficient value in attempting to further reduce uncertainty beyond the point that the constraining factor already provides. This may be because the exact nature of the decision has changed from what it was when the data-to-learning-action process was originally established, or it may be that advances in technology obviate an earlier approach and present an opportunity for learning improvement by reconfiguring the data-to-learning-to-action chain. In such cases, any additional learning capacity beyond the limiting constraint is of no value (and it could be that there is even surplus capacity at the limiting constraint itself, requiring rationalization *below* the capacity level of the limiting constraint). This is essentially a learning-based version of a *stranded asset*.

Determining that there is no value to the extra capacity is also predicated on a determination that there is no other synergy value with respect to the capacity and/or that there is no *option value* for future eventualities. These are possibilities that should obviously be investigated before it is assumed the extra capacity has no value and is rationalized.

As an example of learning-capacity rationalization, we might have the following scenario that is consistent with Figure 8-7. In this scenario, capacity for acquiring and filtering data might have been put in place in anticipation of supporting a decision, and in anticipation that Information Management and Search and Discovery capabilities would likewise be bolstered to support the new data (while the Predictive Analytics and Process and Collaborate elements already had sufficient capabilities to handle the enhanced flows of learning from the upstream elements).

But then something changed. It could be, for example, that a shift in business direction caused the nature of the decision to change such that there was no longer a need for the enhanced learning flow. Therefore, the Data Acquisition and Data Filtering capabilities can no longer be expected to deliver value, and so they should be rationalized in the most cost-effective way possible. On the other hand, if the Predictive Analytics and Process and Collaborate learning capacities have synergy value with other data-to-learning-to-action processes, they, of course, would not necessarily be rationalized, and so their capacity levels might be retained, as is illustrated in Figure 8-7.

As another example, the reverse of the situation that is illustrated in Figure 8-7 could, of course, also occur, whereby there is significant capacity in the Information Management and Search and Discovery elements that was predicated on supporting human-based analysis and decision making. But a new machine learning–based approach might instead put a premium on more data and the filtering of that data, with less need for human-oriented taxonomies and search capabilities, which might therefore now be rationalized.

While building *additional* learning capacity generally comes more naturally, the dynamics of changing markets, new competitive threats, and rapid advances in technology demand that a hard, continuing look at *rationalizing* learning capacity also occurs. This can be the more challenging exercise because it requires overcoming that common cognitive bias of *escalation of commitment* and its resulting inappropriate fixation on sunk costs.

Full Method Overview

With the additional learning throughput strategies outlined in this chapter, we are now in a position to put into practice the full optimizing data-to-learning-to-action method. Here's a summary of the steps, although the exact order and emphasis can vary depending on business realities.

- Determine value drivers and the key decisions related to the highest-leverage value drivers.

- Work backward from the decisions to determine the uncertainties that influence the decisions and model the current learning flows of the data-to-learning-to-action processes that are associated with each of the key uncertainties.

- Identify independent learning flows within each of the data-to-learning-to-action processes.

- Develop learning-value diagrams and identify the learning variables associated with each of the learning flows.

- Determine learning synergies by identifying common learning variables across data-to-learning-to-action processes.

- Identify the limiting constraint of each learning flow.

- Determine potential people/process/technology solutions to the limiting constraints and quantify the expected total value of each solution.

- Tune the value quantifications as required based on synergy, constraint look-ahead, and value-dependency factors.

- Prioritize the solutions by optimizing for expected total value and then implement the solutions based on the prioritization.

- Evaluate results and rerun the process on a continuing basis.

Consistently and broadly applying these steps will set up any organization for continuous business-performance success.

Summary

In this chapter, we reviewed additional learning-throughput strategies, including when to work on resolving learning constraints in parallel, identifying synergies across data-to-learning-to-action processes, and looking ahead to the next sequence of constraints when considering solutions to a limiting constraint of a data-to-learning-to-action process. We discussed a look-ahead strategy with respect to solutions to constraints that require anticipated capabilities that will only be available in the future. We found that an optimization model that maximizes expected total value can lead us to the right solution choices regardless of the number of different options, the complexity of look-ahead, and/or learning-capability granularity. We also covered the reality of situations for which the right strategy is rationalizing learning capacity rather than increasing it. Finally, we outlined the steps of the full optimizing data-to-learning-to-action method, setting us up for a look into common patterns and examples of learning constraints and solutions in the next chapter.

Patterns of Learning Constraints and Solutions

We now have all the tools in place to optimize one or more data-to-learning-to-action processes and, more specifically, to *maximize the value of learning* that is associated with data-to-learning-to-action processes. So, in this chapter, we'll work through some common patterns of learning constraints and potential solutions that are applicable to a wide variety of organizations and functional areas. The solutions invariably rely on people-, process-, or technology-based capabilities and, most typically, combinations of the three. We'll work through these examples by traversing backward along the chain from the targeted decision, which is the sequence that is most appropriate to be applied in any real-world setting. And, of course, that targeted decision should be one that has been determined to have significant leverage on a value driver for the organization. We will spend time on each element of the data-to-learning-to-action chain and discuss some of the common constraints that are associated

© Steven Flinn 2018

S. Flinn, *Optimizing Data-to-Learning-to-Action*,

https://doi.org/10.1007/978-1-4842-3531-7_9

with each element and some typical potential solutions to these constraints. These examples will hopefully resonate with some of the data-to-learning-to-action processes in your own organization and help jumpstart your analysis.

Determining the Value

An initial step of our method is to understand the value potential of improving the decision that is associated with a data-to-learning-to-action process to a level of *perfect* decision making or close to it, as we outlined in Chapter 6. We then determine the *uncertainties* that keep us from attaining perfect decision making and identify those that have the most significant effect, and then we determine the value of perfectly resolving these significant uncertainties. Then, as we systematically work backward along the chain, we seek to identify the factors that are the cause of the uncertainties, categorize the factors by learning flows, identify the limiting constraints within each of the learning flows on learning throughput and value, and then explore the art of the possible in resolving the limiting constraints.

Although we might not be able to realistically achieve perfect decision making, we strive to identify solutions to the limiting learning constraints that can get us closer to perfection, and, importantly, we can determine the value of doing so, which is the *expected learning value* of a solution, as we also reviewed in Chapter 6. This means we also have a quantification of the expected *net* value, i.e., the *expected total value*, of implementing the solution, which can then effectively compete for resources with all the other investment opportunities that an organization is considering, as we discussed in Chapter 7.

And as we reviewed in Chapter 8, we may also need to consider if learning synergy across data-to-learning-to-action processes might influence the expected total value of potential solutions, as well as constraint look-ahead considerations and any auxiliary scenarios that might be required to account for solution granularity and anticipated future capabilities.

Process and Collaborate Constraints

As the data-to-learning-to-action element closest to the decision point, we start our trek upstream along the data-to-learning-to-action chain with the Process and Collaborate element. As we will do in turn as we encounter each of the upstream elements that flow toward the Process and Collaborate element, and as we discussed in Chapter 5, if we identify an uncertainty that would be potentially valuable to reduce (i.e., could influence our targeted decision), we need to segment the causal factors of the current level of uncertainty as follows:

- Causal factors that are attributable to the operations *within* the Process and Collaborate element itself

- Causal factors that are attributable to an insufficiency with respect to the flow of learning *into* the element from the upstream elements

This process of identifying causal factors enables a determination of the limiting constraints of each learning flow, as is illustrated again for convenience in Figure 9-1.

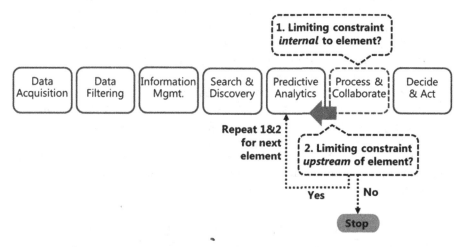

Figure 9-1. Procedure for determining limiting constraints on value in a data-to-learning-to-action chain

The core of our process for identifying key uncertainties and their internal and external causal factors for each of the elements of the chain as we work backward comprises applying a structured interview process during discussions with participants in the specific element of the data-to-learning-to-action chain that we are currently analyzing. We apply a question-based "funneling" approach to home in on the causal factors, supplemented by follow-up analysis and modeling. As we discussed in Chapter 5, it often makes sense to perform this process in two stages, with the first pass primarily oriented toward getting the "lay of the land" of the data-to-learning-to-action process so as to be able to ask the most effective questions in the second pass, as well as to build good working relationships with participants. The following sample interview questions, and similar examples for the other elements of the chain in upcoming sections of this chapter, exemplify this interviewing technique.

For the Process and Collaborate element, we can lead off with questions to participants such as "What are the *factors* that most *limit* your ability to make the best possible decision about *x*?" and "Which of those factors are most important and why?" Then, based upon the answers to those questions, for each of the factors, such as factor *y*, that are indicated to be the most important ones, we follow up with questions such as "How much better would your decisions be if factor *y* were improved to a certain level, *z*?" And then follow up with "What would be required to improve factor *y* to that level?" The responses to this question could result in candidate learning variables to include within learning-value diagrams as the analysis proceeds.

There might be multiple potential improvement factors that could influence factor *y*, and we would continue with similar questions along each of those factor threads. We would want to follow this procedure for as many of the decision-making participants as possible and then cross-calibrate our findings. We might need to go back to some of the participants and do some further interviewing based on the cross-calibration. This could be performed individually, or it might be beneficial to conduct it as a group conversation so that participants can discuss their different assumptions and perspectives, and to determine whether, if the assumptions and perspectives became level-set among the group, there would then be agreement on the relative leverage of the factors. Based upon the results of this process of dialogue with and among participants of the Process and Collaborate element, we can begin developing learning-value diagrams, probabilistic assessments of uncertainties, and associated modeling.

As the most downstream element of the data-to-learning-action chain, it can generally be expected that the highest-leverage learning constraints and solutions of the Process and Collaborate element will be tilted toward the ways people *work together* and *collectively make decisions*. Typical decision-making issues to be on the lookout for as the Process and Collaborate element is analyzed can be divided into the following two categories:

- **Problems with group decision making.** Group decision-making problems may be exacerbated by underlying technology and process constraints, as is the case in our bid-pricing example that we have used throughout the book. But they are even more likely to be primarily the result of problems of interpersonal dynamics, such as fears of being judged or embarrassed, insufficient leadership guidance, less-than-optimal diversity of thinking (i.e., groupthink), escalation of commitment, and/or lack of an established decision-making process or an inadequate one. Additionally, or alternatively, required flows of learning from upstream data-to-learning-to-action elements may not be available or may not be timely.

- **Problems with individual decision making.** Problems with individual decision making can include the well-known inherent shortcomings that we have discussed when people make decisions in the face of significant complexity and uncertainty. But this may be exacerbated by insufficient knowledge and skills with respect to the subject of the decisions, the inability to tap into expertise as it is needed, or having insufficient time and attention to devote to the decisions. As in the case of group decisions, required data and information from upstream data-to-learning-to-action elements may not be available or may not be timely.

Given that the constraints in the Process and Collaborate element are typically more directly people-related, for those constraints that are attributable to the operations *within* the element, appropriate *coaching* and *education* will be fundamental to most solutions. The coaching and education will typically fall into the following categories:

- Leadership and team building

- Awareness of cognitive barriers that inhibit optimal individual and group decision making and methods to overcome the barriers[1]

- Subject matter skill development

It is useful to then top off that education with an overview of the optimizing data-to-learning-to-action method so that everyone who needs to understand the context of, and motivation for, these educational-based solutions is provided with that understanding.

And from a technology perspective, it is often the case that the interactions among decision makers in the Process and Collaborate element are done through the use of general-purpose collaborative platforms that include a mix of communications modes, such as email, messaging, video conferencing, social networking, group management, and perhaps project or workflow management. Better *integration* of these different modes of collaboration may be the path to solving constraints in some cases. And as the technology continues to progress, there may be opportunities to apply more exotic tools such as immersive environments, including virtual and mixed reality, to facilitate valuable engagement among decision makers.

---

[1]For techniques to overcome common cognitive barriers to individual and group decision making, see again: Bang, Dan, and Chris Frith, "Making better decisions in groups", *Royal Society Open Science*, August 2017. http://rsos.royalsocietypublishing.org/content/4/8/170193

To the extent that it is found that the technology that is being used to make collaborative decisions contributes to a limiting constraint, potential technology-based solutions are typically going to be driven by the *synergy value* across multiple data-to-learning-to-action processes, as illustrated by Figure 9-2. That's because it is most typically the case that the loss in value that is directly attributable to technology-related constraints in the Process and Collaborate element for *any given* data-to-learning-to-action process is not sufficient to justify a significant technology upgrade. However, carefully considering the value of the potential upgrade *across* multiple data-to-learning-to-action processes may demonstrate just such a justification.

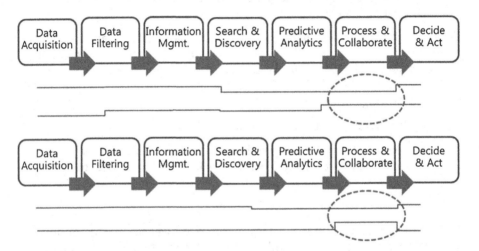

Figure 9-2. Identifying synergy value across mulitple data-to-learning-to-action processes for Process and Collaborate support systems

For the Process aspect of the Process and Collaborate element, methods of delivering the right information that is also structured the best possible way to support decision makers, such as through business intelligence applications and associated dashboards of key and timely *metrics*, may serve to help de-constrain learning. Of course, determining what metrics are most important for the decisions that are being made is exactly what our learning constraint and solution identification process is all about, and many of the key metrics will therefore necessarily directly fall out of our process (more on that in the next chapter).

Whether it is coaching, education, process enhancements, and/or technology-based solutions that are determined to be a part of the learning constraint solution mix, the advantage with the data-to-learning-to-action approach is that we have a robust and credible way to quantify the value of these solutions. This is particularly important for the Process and Collaborate element, which in many organizations has traditionally tended to get shorted on supporting education and infrastructure, perhaps in part because the nature of the element has generally relegated the associated benefits analysis to merely reciting anecdotes and "feel-good" appeals. Without the benefit of the optimizing data-to-learning-to-action method, the value of people-based "soft" investments has often simply seemed too slippery to quantify, while platforms that support collaboration have often been perceived to have such diffuse benefits that it is assumed to be fruitless to rigorously quantify their net value-add. But, we can now certainly do better than that!

It should also be remembered that as we work through the streams of learning and associated constraints in this or any of the other elements of the chain that we need to be careful to identify which streams constitute *independent flows* and associated constraints so that there is the option to address them in parallel if the value modeling suggests that it would be appropriate to do so.

It may be that for a given data-to-learning-to-action process all the major learning constraints are directly attributable to the internal operations of the Process and Collaborate element. If so, and in accordance with our procedure of Figure 9-1, we need not work any further upstream in the data-to-learning-to-action process chain. But if that is not the case, we follow upstream the learning flow(s) that input into the Process and Collaborate element, starting with the Predictive Analytics element.

Predictive Analytics Constraints

The forward flow of learning from the Predictive Analytics element can potentially support the associated data-to-learning-to-action process in two basic ways. It can deliver predictive insights downstream to the Process and Collaborate element to facilitate *human-based* decisions, or it can deliver predictions that enable *automated* decisions, essentially by-passing the Process and Collaborate element. Of course, the scope of the latter approach increasingly expands as software continues to inevitably "eat the world" and AI, in turn, continues to "eat software."[2] And, moreover, with the rise of the field of data science and the data scientist role, and in combination with literally monthly advances in automated capabilities for predictive analytics, this element is truly undergoing a revolution generally—and if not already, likely soon for your organization, as well.

---

[2]Andreessen, Mark, "Why Software Is Eating the World", *Wall Street Journal*, August 20, 2011, https://www.wsj.com/articles/SB10001424053111903480904576512250915629460; Simonite, Tom, "Nvidia CEO: Software Is Eating the World, but AI Is Going to Eat Software", *MIT Technology Review*, May 12, 2017, https://www.technologyreview.com/s/607831/nvidia-ceo-software-is-eating-the-world-but-ai-is-going-to-eat-software/

For example, *self-learning* AI is already beginning to rapidly "eat" *directly programmed* AI, which itself was novel just a few short years ago in some areas, and has not even yet been applied in other areas! A watershed example is AlphaZero's ability to *self-learn* games of skill such as chess from basic principles within a very short time, and then exceed not only human players' abilities, but also the abilities of the best *human-programmed* AI that has evolved over the past twenty years.[3] This progression, as illustrated in Figure 9-3 for the field of chess, will no doubt be recapitulated in many predictive-analytics applications in the enterprise and will surely have profound effects in many areas.

Eras of Chess Playing Dominance

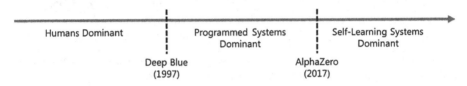

Figure 9-3. AI progression in the field of chess

At a minimum, in the near term these advances in AI/machine learning constitute one of the critical areas of anticipated capabilities that will need to be carefully monitored for potential application in any given data-to-learning-to-action optimization. In Chapter 4, we reviewed some of the major enterprise applications and how their various features mapped to data-to-learning-to-action elements, and a common theme was that many of these applications were extending their feature sets into the Predictive Analytics element as machine learning–based capabilities continued to advance. Although we don't yet know exactly for what applications they will best fit, the advent of *self-learning systems* will surely extend that trend, and therefore will demand continuously monitoring and evaluating for optimizing data-to-learning-to-action purposes.

As with each of our element-specific analyses, we segment the constraints on learning into those constraints that are *internal* to the operations of the Predictive Analytics element itself and those constraints that are attributable to *inputs into* the element. Predictive analytics boils down to the following:

Data/information + Models/algorithms ==> Predictions

---

[3] Silver, David et al., "Mastering Chess and Shogi by Self-Play with a General Reinforcement Learning Algorithm", *Cornell University Library Arxiv.org*, December 5, 2017. https://arxiv.org/pdf/1712.01815.pdf

This basic fact makes for a clear-cut segmentation: if the constraint is internal, it is typically a *predictive modeling–based* issue (whether that modeling occurs in mind or machine), and if the constraint is the result of a deficiency of an input into the element, it is a *data- or information-based* issue (including a possible information *delivery* problem due to a deficiency of the Search and Discovery element).

As was the case for the Process and Collaborate element, we can apply a structured questioning process to assist in identifying both internal and external constraints. A good general question to pose before following up on specific constraints identified by the Process and Collaborate analysis to gain that first-pass "lay of the land," and that is relevant not only for this element, but also as we address any of the elements upstream of it, relates to the *effort* and *attention priorities* of participants in the Predictive Analytics element. Their responses can provide insights that can help us determine how aligned the participants are with the first-cut value models that we have developed so far in our analysis. For example, we might ask "On what problems or models do you spend most of your time, and why?" And then follow up on their answers to understand more details as required.

Then, we need to proceed to the core of the interviewing, which is following up on the constraints identified in the Process and Collaborate element that we found were a result of inputs, or the lack thereof, from the Predictive Analytics element. For constraints that are identified as internal to the Predictive Analytics element, we might ask: "What is the theoretical best predictive capabilities with respect to uncertainty x given the data and information that is currently available and is being used as a basis for the predictions?" And then follow up with "What predictive processes, models, and/or skills would enable achieving as close to that theoretical best as possible?" There might be multiple possibilities posited by the interviewees, and we would then follow each of those threads to learn more. To understand external-origin constraints, we might ask: "What data, enhanced filtering of data, or information-delivery means would enable your predictions of x to be even better, and how much better?" Once again, we might have multiple possibilities to follow up on with additional questioning. In some cases, it might be beneficial to get perspectives from outside experts to ensure that all possibilities are considered.

As is always the case after an initial round of interviews, we might then need to go back to some of the participants and conduct further questioning based on a cross-calibration of what we learned from the interviews performed to date. And, again, in some circumstances, particularly when the group dynamics are not favorable to obtaining fully candid perspectives from the interviewees in group discussions, this follow-up could be conducted individually with interviewees (and if this needs to be the case, we might have stumbled onto a learning constraint!). Or if the group dynamics are healthy, it might be beneficial to cross-calibrate with groups of participants so that their different assumptions and perspectives can be discussed and a determination be made as to whether, if those assumptions and perspectives become commonly understood among the group, there is agreement on relative leverage of constraint factors, levels of uncertainties, and potential solution ideas. As the information we receive becomes more solid, we can build out our models of the most important constraints, learning variables, probabilistic assessments, and potential solution ideas.

The following is a sample of common areas of uncertainty by functional area for which the application of enhanced predictive analytics to reduce the uncertainty can potentially have high-value leverage:

- **R&D:** performance attributes of new products, time-to-market for new products, cost of new product development

- **Marketing:** qualified leads generated by a marketing campaign, costs per qualified lead, percentage of closed deals from qualified leads, price elasticity of demand, performance attributes, and market timing of competitor products

- **Sales:** probability of closing a deal for a given qualified lead, expected negotiated prices, customer-retention rate, customer-support costs, lifetime value of a customer

- **Supply Chain:** probability of supply disruption at each step in the supply chain, product demand over time, variance of product demand, level of inventory write-offs

- **Human Resources:** performance levels of recruits, performance of promoted employees, future required employee compensation levels, expected level of employee morale, employee retention rate

Of course, the key is to understand the *factors* behind these types of uncertainties and identify the constraints and associated learning variables for the most important of the factors. In identifying internal, model-based constraints, we need to be careful that during the interview process and subsequent encoding of the results into our modeling that we fully consider people, process, and technology factors. Some common causes of learning constraints *within* the Predictive Analytics element include the following:

- Resources and/or processes oriented to other than the most important problems—that is, *not aligned* with the *highest-leverage* decisions and quantifiable value

- Lack of qualified people *resources* or *skills* such as data-science capabilities

- Lack of adequate *systems-based tools* to support the people resources

- Failure to effectively *apply* the available data and information sources

- Failure to effectively *deliver* predictive analytics results to the other elements of the data-to-learning-to-action process

Most of these problems, except for the last item, which is a failure to deliver results to other elements, are fundamentally about wringing the greatest possible *predictive power* out of the data and information that are available. Some of the general potential solutions and considerations for these types of issues include the following.

- Use the optimizing data-to-learning-to-action method that is outlined by this book to ensure that scarce data and decision-science resources are working on the areas for which the expected value is *greatest*!

- *Acquire* expertise if required, but only after you know exactly the intended applications of that expertise and the expected value of that expertise by, again, applying the approach in this book!

- Provide *targeted* training and education resources that are aligned with areas that require the most attention based on expected total value (yes, again, by applying the method advocated by this book!)

- Ensure that the *processes* for generating predictive analytics are aligned with the desired results, which means that they are particularly focused on the learning variables that have greatest value leverage.

- Determine if the availability of *enhanced modeling tools and applications* can improve predictive power. It is particularly important to monitor the availability of cloud-based capabilities because they are evolving rapidly and literally advancing daily. Machine learning–based technology is also rapidly advancing and may be applicable for some types of predictive analytics–based problems and so should be continuously monitored. Of course, cut through the hype by forcing suppliers of such technology to be as *quantitative* as possible with respect to the potential reductions in uncertainty the technology can be expected to deliver so that you can assess expected value—don't accept hand waving!

- Given that we are still in the early stages of the lifecycle of some predictive-analytics approaches, such as machine learning–based applications, there may be a build-versus-buy sub-decision to consider. Prototyping with the expectation of throwing the prototype away in anticipation of rapidly advancing commercial capabilities may be warranted in some situations.

- Ensure opportunities are considered for not only delivering predictive capabilities that will be *directly* applied by decision makers, but also for capabilities that *embed* predictive inferences within the Information Management and Search and Discovery elements, with the understanding, however, that realistically most of these types of capabilities will likely be best provided within commercial packages.

These are some general solution ideas to consider. But as is the case with other elements, it may take some *creative thinking* to wring more *predictive power* out of the data and information that is already *available*, or to determine *new* data and information inputs into the Predictive Analytics element that could make a difference, but that have not been seriously considered previously. It should never be confused that the additional process structure that the optimizing data-to-learning-to-action approach advocates is suggesting that creativity and innovation are not necessary co-ingredients!

As with our analysis of each of the elements of the data-to-learning-to-action process, we should identify independent learning flows and associated constraints and evaluate the merit of de-constraining them in parallel across the independent streams. We should also be particularly mindful of looking ahead to *anticipated* technology advances for the Predictive Analytics element when evaluating de-constraining approaches given the rapid advances in fields such as machine learning. And, of course, for expertise and technology that will apply to multiple data-to-learning-to-action processes we need to be careful to calculate the synergy value by summing up the expected value-add contributions to all the relevant data-to-learning-to-action processes.

If we find that the limiting constraints of the independent learning flows of a data-to-learning-to-action process are directly attributable to the *internal operations* of the Predictive Analytics element, we do not have to investigate the inputs from the upstream elements and can go ahead and dive into assessing the expected value of potential constraint solutions such as the examples previously listed. But if that is not the case, we follow upstream the associated learning flow(s) that input into the Predictive Analytics element, which will generally be via the Search and Discovery and Information Management elements.

Search and Discovery Constraints

With Search and Discovery, we are venturing further back in the data-to-learning-to-action chain in the hunt for the limiting constraints on value. Remember—we may not get this far, and *we don't want to get this far* unless we need to! The limiting constraint for any given learning flow may well be within an element that is downstream from Search and Discovery, and, if so, there is simply *no reason* to do any work upstream of that constraint. That's the whole point of the optimizing data-to-learning-to-action approach—it's not about "boiling the ocean." It's having a laser-like focus on the limiting constraint on value and finding solutions that serve to release that constrained value. Having said that, at some point, for some of the data-to-learning-to-action processes in an organization, Search and Discovery will inevitably embody a limiting constraint that requires addressing.

Search and discovery functionality is, of course, primarily directly used by individual users, delivering information to them that serves as an input to their activities in the Predictive Analytics and Process and Collaborate elements of the data-to-learning-to-action chain. As we discussed in Chapter 3, search delivers information based on *explicit* information that is provided by the user, while discovery is more generalized and delivers information that is in accordance with other, or additional, inferences that are derived from current context and/or historical behaviors.

The constraints in both search and discovery basically boil down to a familiar type 1 and type 2 categorization of contrasting errors: first, erroneously delivering information that is not relevant and that we *don't need*, and hence wasting our attention and time, and second, erroneously *not* delivering the information that we *do need*, which is an even worse outcome than wasting our time. And as we analyze these two types of learning constraints, as usual, we segment the constraints by those that are *internal* to the operations of the Search and Discovery element and those constraints that are attributable to the *inputs into* the element, typically inputs that flow from the Information Management element of the data-to-learning-to-action chain.

As in the case of Predictive Analytics, a core aspect of Search and Discovery is "data/information plus models/algorithms." And this fact again makes for a clear-cut segmentation: if the constraint is internal, it is typically an algorithmic-based issue, and if the constraint is the result of a deficiency of an input to the element, it is a data- or information-based issue. However, since search and discovery are functions that are used directly by *people*, there is another important internal factor besides algorithms that should be considered as a possible contributor to learning constraints: the system's interface with the user.

As with the previous elements, we can apply a structured interviewing process to facilitate identifying both internal and external search and discovery constraints. And our interviews should be focused on asking questions that relate specifically to the inputs from Search and Discovery or the lack thereof—let's label these constraints *x*—that were found to be constraining the participants of the Predictive Analytics or the Process and Collaborate elements. Our questioning should therefore begin with these *users* of the search and discovery functions.

For search, a first fundamental question to ask users is: "Are you able to effectively search for and find information related to *x*?" And then: "Why or why not?" We then follow the answers to the "why not"s deeper, identifying the contributing factors as we go, evaluating their relative effect, prioritizing the factors, and shaping our continuing questioning and analysis as we go along based upon our evaluations and prioritizations. And, as usual, we would typically follow up with participants after an initial round of interviews to ensure mutual understanding and to cross-calibrate on the identified constraints and their relative importance.

Constraints that are likely to be found in these exercises include the following:

- The search system often fails to *understand* users' queries and/or their context (problem *internal* to the element).

- It takes *too long* to find what users are searching for; i.e., search doesn't immediately deliver the *most relevant* information (problem *internal* to the element).

- The search system has limited understanding of the *semantics* of the content that is being searched (problem *internal* to the element).

- The search system does not effectively handle *non-text* sources of information such as pictorial or audio-based information (problem *internal* to the element).

- Users do not fully trust the results of their search requests (problem may be *internal and/or external* to the element).

- The information that users are looking for is simply *not available* for the search function to find (*input problem* from an upstream element).

Similarly, for discovery functionality, a fundamental question to ask users is: "Does relevant information related to *x* automatically get surfaced to you? If not, is it the case that you are receiving information about *x*, but that it isn't relevant? Or is it the case that you are not receiving automatically *any* information about *x*?" The reality is that for many organizations there is currently limited discovery functionality in place, so in the near term we are more apt to get a confirmation of the second option, but that is changing rapidly as machine learning–based capabilities continue to advance.

Typical constraints that will be revealed from interviews regarding discovery are as follows:

- Discovery functionality is *not available* to users for the relevant use case (problem *internal* to the element).

- Discovery functionality *is* available to users, but it's not *good enough*—it doesn't surface what is most personally and contextually relevant and/or surfaces too much information that is not relevant (problem *internal* to the element).

- Discovery functionality *is* available, but the system's suggestions can be inscrutable, thereby diminishing users' trust in the discovery results, and the system does not provide users with adequate *explanations* for the suggestions.

- Discovery functionality *is* available, but it does not have *access* to the information that is most relevant to users (may be *combination internal/input* problem).

- Discovery functionality is available, but the relevant information is *not available within the organization* for the discovery function to surface (*input problem* from an upstream element).

Some general potential solutions to be considered for constraints that are *internal* to the Search and Discovery element include:

- Search functionality that more accurately infers the *meaning* of user search requests

- Enhanced discovery *contextualization* capabilities that are more effective at surfacing information that is relevant to the *current activity* of the user

- Search and discovery functionality that is accessible through an *interface* that enables a more effective and efficient understanding of the users' current context and intent

- Search and discovery functionality that is accessible through an intelligent, *conversational-based* interface

- Enhanced search and discovery *personalization* capabilities that are more effective at *anticipating* and surfacing information that is *personally relevant* to the user based upon inferred user interests and/or expertise

- Search and discovery functionality that better *understands* the *context or meaning* of the collection of content that is being searched or considered for surfacing

- Search and discovery functionality that provides meaningful *explanations* to users for search results, as well as for recommendations and suggestions, thereby enhancing users' trust in the system as well as providing useful auxiliary information

While the preceding potential solutions have been categorized as being internal to the Search and Discovery element, some may require enhancements to be implemented within the Information Management element as well. For example, for enhanced search and discovery personalization, it may be that a richer set of *behavioral information* needs to be available in the Information Management element. And for more effective understanding of the context and meaning of documents that are being searched, general-purpose and/ or domain-specific ontologies may need to be available in the Information Management element for search and discovery functions to access. This, of course, may require specialized technology and/or ontology skills be applied. Further, for some needs, specialized data warehouses may be required for search and discovery to operate against. Of course, as in many other areas, neural network–based approaches may also become increasingly applicable for inferring the semantics of content in the enterprise context.

Remember too that there is a time element to consider—the learning *throughput* that we want to optimize is a *rate*, actionable learning per unit of time. It is usually not enough to just *eventually* get the information that is required to reduce uncertainty, and to therefore be able to more effectively predict. So, search and discovery that takes too long to surface the relevant information to a user is a drain on value. And for many data-to-learning-to-action processes, that is not just a generalized, productivity-based concern— delays in learning that lead to delays in taking action can *directly* cause measurable foregone value (such as in bidding situations).

As is the case when analyzing learning flows within each of the elements of the data-to-learning-to-action process, we should identify Search and Discovery–related flows that are independent and identify the associated limiting constraints of each of the independent flows, and then evaluate the merit of de-constraining the independent flows in parallel. And we should be mindful of the reality that search and discovery functionality is primarily going to be supplied by commercial packages, and the justification for the enhanced features of the packages will most likely be driven by the *synergy value* of multiple data-to-learning-to-action processes.

If we find that the limiting constraint on value is directly attributable to the *internal operations* of the Search and Discovery element, we need not investigate the inputs from the elements upstream of it, and we can therefore proceed to assessing the expected total value of potential solutions to the limiting constraints, such as the examples listed previously in this section. But if that is not the case, we follow upstream the constraint that is limiting the value of the Search and Discovery element, and the next stop in identifying that culprit is the Information Management element!

Information Management Constraints

We have now arrived at the Information Management element, and we should be here only because we found from our previous investigations that the limiting constraint of a specific learning flow was *not internal* to the Process and Collaborate element, was *not internal* to the Predictive Analytics element, and was *not internal* to the Search and Discovery element, and so the constraint must be further upstream in the data-to-learning-to-action chain. So, we have gotten here only grudgingly, which is exactly the way it should be. Too often organizations jump to a technology solution without understanding thoroughly where the *real* problem lies. Our step-wise process of working backward from decisions serves to overcome that inclination.

For the Information Management element, the constraints on learning will fall into two categories:

- The required information to overcome the constraint exists within the organization but is *not readily accessible* to users who need the information (problem *internal* to the element).

- The required information to overcome the constraint *does not currently exist* anywhere within the organization (*input problem* from an upstream element).

If it is the latter issue, we need to continue our swim upstream to investigate the Data Filtering element. If it is the former, we have some more analysis to do within the Information Management element, as follows.

As usual, we begin our analysis of this element by interviewing the people with relevant knowledge and expertise, which in this case may include enterprise architects, knowledge managers, taxonomists/ontologists (assuming these roles exist in the organization), and database administrators, and then we follow the conversation threads to identify the specific factors that are the cause of the constraints.

Constraints that are internal to the Information Management element and are likely to emerge from the conversations include the following:

- The required information is accessible by *some* members of the organization but not *all* members who need it because it is *tacit knowledge* currently residing only in the minds of some members of the organization.

- The required information is accessible by *some* users but not *all* users who need it because of *technology incompatibilities*.

- The required information is accessible by some users but not all the users who it need because of *privacy or security restrictions*.

- The required information is not readily accessible because it is not *properly structured for the use cases* that have been determined in the downstream analyses to be constrained due to the inability to readily access the required information. This condition may apply not only for human users, but also for system-based access and consumption of information, such as for the training of neural networks.

The following are sample potential solutions, comprising technological and/or skill-based approaches, to the preceding internal constraints:

- Determine if tacit knowledge can be *embodied in computer-based formats or structures* that enable wider access of the information within the organization. As might be expected, the bias of this book is that this should be increasingly feasible for most information, even if it wasn't in the past, given the advances of technology that, for example, can effectively interpret and search video and audio formats.

- Explore overcoming technology incompatibilities by evaluating the opportunity to *consolidate* information-management platforms or to apply *connectors* that integrate disparate technologies.

- Apply *architecture of learning*–type structures as discussed in Chapter 3 that encode behavioral information, inferences of interests, and expertise from the behavioral information to enable adaptive personalization of search and discovery functions.

- For privacy or security restrictions, ensure that the policies are sensible in view of a *realistic appraisal of costs and benefits* (and the good news is that you will now actually have quantification of the benefits in hand) with respect to the specific users who need the information to overcome a limiting constraint.

- Acquire appropriate database, taxonomy, ontology, and data-science skills and/or technology as required to better structure and organize information to effectively support the constraining use cases. In some cases, more specialized structures such as NoSQL structures or data warehouses may be warranted. Applying graph-based structures may be an avenue for synergizing what are otherwise disparate pools of information.

Of course, for cases in which the required information that is causing a limiting constraint does not currently exist anywhere within the organization, we must again look upstream for a potential solution, so we next turn to the Data Filtering element.

Data Filtering Constraints

If we have gotten this far, to the Data Filtering element, it is most likely that the information that is not available to the Information Management element is because the underlying data simply does not exist within the organization, in which case we need to proceed one more step upstream to the Data Acquisition element of the data-to-learning-to-action chain.

However, in some cases, the raw data that is required *is* available within an organization, but it has not been sufficiently filtered such that it is transformed into usable information, which, of course, is the job of the Data Filtering element. As always, we first interview the potential users of the information to better understand what is constraining their relevant use cases, as well as interview those with relevant skills, such as data scientists, who can provide insights into root causes and potential solutions.

Some common data filtering–related constraints that we are likely to find from our interviewing include the following:

- The raw data contains orders of magnitude more *extraneous* data than the useful data that represents the *information signal* that is required by the key use cases. Therefore, the data in its current, raw state is unusable.

- The raw data includes a significant number of *spurious or outlier data* that make it unusable.

- The data needs to be *coupled with other data* to be usable (e.g., event data needs to be timestamped).

- The data needs to be *translated* into a different format or language to be usable.

Example potential solutions (which may need to be applied in combination) to these constraining factors include the following:

- Apply technology to extract the *relevant* information from the vast corpus of data. For example, neural network–based technology might be applied to video streams to extract the specific features of interest from the massive number of mostly extraneous pixels in the video streams. Another example is the "waste makes haste" approach of high-throughput experimentation discussed in Chapter 4 in which the vast majority of experimental results need to be rapidly filtered from further consideration so that the results that could really matter receive appropriate focus.

- Apply *data-science* skills to condition data by removing spurious and outlier data points so that the data is fit for use.

- *Merge* or *augment* data sets as required to create usable information. For example, matching individuals to behavioral data or timestamping event data.

- Apply ever-improving *translation or interpretive* technology to convert documents and other content such as pictures into usable information for specific users and their use cases.

Of course, if the raw data that is required is not available at all within the organization, then we move on to investigate the initial element of the data-to-learning-to-action process, Data Acquisition.

Data Acquisition Constraints

We're now at the end of the chain, so the learning-value buck stops here! Any limiting constraint we find now is necessarily *internal* to the Data Acquisition element, as illustrated by Figure 9-4, so that buck can't be passed. It must be the case that the data that we need is simply not available within the organization, and we therefore need to try to get it. Of course, we need to accept that attaining the desired data simply may not be fully possible. But this is an opportunity to apply an organization's creative powers to do so! If it can be done such that the *expected total value is positive*, then we should, of course, go ahead and do what it takes to attain the data.

Figure 9-4. The learning-value buck stops at the Data Acquisition element

There are, of course, many different types of data and sources of the data that could be considered as solutions to a limiting constraint, so we can only cover a tiny fraction of solution possibilities, but following are some common ones:

- The internet of things and sensor technologies continue to advance and provide the opportunity for capturing operational data from a wide variety of industrial processes and environmental conditions that would have previously been impossible.

- Combinatorial-chemistry and high-throughput experimentation–based approaches can provide additional, perhaps orders of magnitude more, raw data for physical product–based R&D.

- A/B testing can provide new, empirically-based data to help refine user interfaces to maximize customer satisfaction and revenues.

- Increasingly finer-grained consumer behavioral data can be captured or acquired, including purchasing histories, location histories, viewing and listening habits, social interactions, and so on.

- Employee behavioral data can be captured by, and accessed from, the latest generation of enterprise applications and platforms.

- And while the preceding technology-related opportunities are important to consider, we shouldn't lose sight of the fact that a significant source of untapped data resides within the minds of the members of an organization, and conversations and interviewing processes can be applied to capture that data!

It is possible, of course, that we find that we are not able to acquire the data that we need in order to relax the limiting constraint, at least in the near term, or to do so cost-effectively. But as with any other limiting constraint that exists anywhere in the data-to-learning-to-action chain, because we have a quantified cost of the limiting constraint in hand, we can now be continually on the lookout for viable solution options as technology progresses and costs decrease, and be ready to take advantage of cost-effective advances as soon as they become available.

Summary

In this chapter, we worked through the optimizing data-to-learning-to-action process and touched on some of the typical learning constraints that are encountered, as well as some of the potential solutions to those constraints. The general steps for optimizing a data-to-learning-to-action process that we covered are as follows:

- Quantify the value of *perfect* decision making with respect to the primary decision associated with the data-to-learning-to-action process to create a baseline upper-bound value for resolving learning bottlenecks that are constraining value.

- Identify the significant uncertainties that are causes of the less-than-perfect decision making and quantify the value of *perfectly* resolving each of these high-leverage uncertainties.

- Begin the analysis of the *learning constraints* that are the cause of the uncertainties by working backward to the element of the data-to-learning-to-action process that is closest to the decision, the Process and Collaborate element.

- Identify learning flows and limiting constraints associated with the uncertainties that are *independent* of one another.

- Determine if each of the independent limiting constraints are *internal* to the Process and Collaborate element or are associated with a required *input* from an upstream element.

- For limiting constraints that are internal, identify potential *solutions* and determine preliminary *expected total values* of each of the solutions.

- For limiting constraints that are associated with a required input from an element further upstream of the data-to-learning-to-action chain, begin analyzing the limiting constraints in that upstream element *by applying the same process* as was performed in the previously analyzed element.

- Continue this process until *all* the limiting constraints that have been identified are determined to be internal to an element of the data-to-learning-to-action chain, and this condition must necessarily be satisfied if we reach the furthest upstream element, Data Acquisition.

- The expected total value of some solutions will be dependent on the *synergy value* of addressing limiting constraints across multiple data-to-learning-to-action chains.

- Implement the solutions that have an *expected total value that is positive*, and *monitor* and *learn* from the results.

These steps are further summarized in diagrammatic form by Figure 9-5. The primary steps begin by determining the target decision and the associated data-to-learning-to-action chain from value-driver analysis (1). The most significant uncertainties that influence the target decision are identified and the values of perfectly resolving the uncertainties are estimated (2).

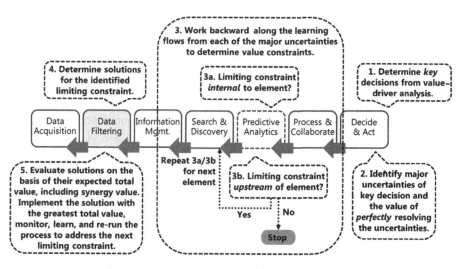

Figure 9-5. Diagrammatic summarization of the optimizing data-to-learning-to-action process

The process then begins working backward along the data-to-learning-to-action chain for each independent flow of learning that corresponds with each of the uncertainties that has significant leverage on the target decision. An iteration step (3) is performed whereby each element of the data-to-learning-to-action chain is examined in turn to determine if there is a limiting constraint to the flow of learning that is internal to the element (3a) and/or if there is a limiting constraint upstream of the element (3b). If the latter is the case, the iterative process moves upstream to the next element. The iteration step necessarily ultimately identifies at least one limiting constraint—in this case, it is premised to be within the Data Filtering element, as is indicated by the element's shading.

People-, process-, and/or technology-based solutions are then identified to address the limiting constraint (4). The solutions are evaluated based on their expected total value, which may include synergy value across data-to-learning-to-action processes, as well as considerations for factors such as the anticipated availability of future capabilities and/or solution granularity. The solution with the greatest positive total value is implemented, the results are monitored, and the resulting learnings are assimilated into the process (5). The process is then re-run to identify and address the next limiting constraints.

That's it! This process can apply in any organization and for any data-to-learning-to-action process. And after getting started applying the method, patterns across an organization's functions and processes soon emerge, and it becomes easier and easier to identify constraints and quantify value. On the other hand, as low-hanging fruit is harvested by applying the method, it should be expected that more and more creativity is required to find cost-effective solutions for the next level of limiting constraints.

While the optimizing data-to-learning-to-action method can seem like more work than status quo processes, it really isn't. It's just a matter of orienting an organization's attention and resources toward what truly matters and not wasting time on what simply doesn't matter. It can be analogized to chess playing. Weaker players or systems tend to look at just about all possible moves in a position, and therefore cannot go very deep on any one of the possible lines of move sequences. Strong players or systems don't waste time on most of the huge number of possible moves. They continuously prune the tree of moves severely, but then they go very deep on the most promising lines.[4]

---

[4]This is the approach of the *self-learning* AlphaZero system, for example, that recently defeated the best *directly programmed* systems: "AlphaZero compensates for the lower number of evaluations by using its deep neural network to focus much more selectively on the most promising variations—arguably a more 'human-like' approach to search ..." Silver, David et al., "Mastering Chess and Shogi by Self-Play with a General Reinforcement Learning Algorithm", *Cornell University Library Arxiv.org*, December 5, 2017. https://arxiv.org/pdf/1712.01815.pdf

Similarly, in the optimizing data-to-learning-to-action approach, we simply don't work on things that are not important. If the upside of reducing an uncertainly is quite low (and often this may not be *intuitively obvious*), we don't work on it. And if it's not a *limiting* constraint, we don't work on it. You will be amazed when the discipline of the optimizing data-to-learning-to-action process is brought to bear how much light it will shed on all the time and money that are being spent on things that don't really matter. And, of course, all that time and money can then be reallocated to dive deep on the things that the optimizing data-to-learning-to-action method demonstrates really do matter!

Organizing for Data-to-Learning-to-Action Success

We've covered a lot of material up to this point, and some of it was perhaps new in concept and maybe even took a couple of passes to fully absorb. But that's the easy part! Now comes what is potentially the harder part and certainly the more important part—making optimizing data-to-learning-to-action an *ongoing* reality in *your* organization!

In this chapter, we will discuss how to get to that ongoing reality. We will cover topics such as scoping the initiative, gaining executive buy-in for the initiative, options for organizing both individual optimizing data-to-learning-to-action projects and broader, company-wide initiatives, and specific change-management issues related to optimizing data-to-learning-to-action initiatives, particularly the what's-in-it-for-me aspects for different parts of the organization. As with any new approach, there will inevitably be skeptics, and

© Steven Flinn 2018
S. Flinn, *Optimizing Data-to-Learning-to-Action*,
https://doi.org/10.1007/978-1-4842-3531-7_10

we will discuss ways to transform those skeptics into valuable resources who will help drive your initiative to success!

Scoping the Project

The guiding principle for scoping an initial optimizing data-to-learning-to-action project is to focus on a decision area that

- is important to your organization;

- is commonly understood to need improvement; and

- is bounded in scope such that the required analysis can be successfully conducted with the resources that are available.

The importance factor can be confirmed (or perhaps disconfirmed) by value-driver analysis. In fact, because value-driver analysis is relatively easy to perform, it is worthwhile to expand that analysis *beyond* just confirming the economic leverage of the candidate decision area being considered. The value-driver analysis will be a valuable piece of work in its own right and will serve to underpin the future optimizing of data-to-learning-to-action processes. And if the importance of the candidate decision area is *disconfirmed* by the value-driver analysis, then there is an opportunity to instead target a decision that appears to have higher-value leverage as evidenced by the value-driver analysis.

The commonly-understood-to-need-improvement factor of a candidate decision area is another case of an intuition that should be verified. A few questions posed to the decision makers should be sufficient for verification purposes. These are questions of the type that will be asked during the analysis anyway, such as "What are the factors that you can think of that seem to be keeping you from making the best decisions possible?" and "Can you think of, off-hand, any ways that those factors could be overcome?" Most likely the answers to these questions will form a basis for further analysis of this decision area. But if you find that the current decisions seem to already be about as good as they could possibly be, it is time to turn to another candidate decision area with more improvement potential.

If the candidate decision area has passed these first two filters, the final consideration is whether you have the *resources* to take on the project. For even the *smallest* optimizing data-to-learning-to-action project, it is recommended that you have at least two resources on the project who have a good grasp of the overall concepts in this book. That enables you to have two people to conduct each interview with process participants, which helps ensure important follow-up questions are not missed. You also need at least one of the resources to be able to handle the modeling required, including the translation of encoded uncertainties into quantified values of learning.

I find that knowledge of the subject matter of the decision area is somewhat of a double-edged sword. On the one hand, *some* knowledge can accelerate the interviewing and subsequent modeling. On the other hand, having little knowledge of the subject matter can lead to asking very useful "naïve" questions that might not otherwise be asked, which in turn can lead to insights that might not otherwise be uncovered. For that reason, it is often beneficial to have at least one resource on the team who does not have much experience with the decision area. More details on organizing and resourcing the project are covered in the section on that topic later in this chapter.

So, if the decision area has high leverage on business performance and has room for improvement, and you can organize the resources needed to successfully apply the optimizing data-to-learning-to-action process, it's time to move to the next step. In most cases, that will be to gain executive buy-in. Even if you thought you could just skunkworks the project, in most cases you will want to get explicit executive buy-in so that you will have fully attentive and cooperative interviewees and implementers, colleagues who will often span multiple organizations and associated executive jurisdictions (e.g., marketing *and* IT).

Gaining Executive Buy-in

Of course, a good first step in getting the executives of your organization on board is to get this book in the hands of your executive teams! Whether line business executives, financial executives, or IT executives, the approach advocated by the book and the specific examples should resonate with executives whatever their area of responsibility.

However, the reality is that only a minority of people will "get it" just from reading this or any other management book and are then ready to charge forth. And even if they *are* ready to move forward, executives are a "show me" bunch—they want to see some results, and they want to see them fairly quickly. So, demonstrating some early value is key. That's why starting with a target decision or two that really matter is important. And then it is important to be able to show some early results that matter by phasing the project around deliverable milestones that will get and keep executives' attention. This is really just the proof of concept or pilot approach that we discussed with respect to the overall business-renewal lifecycle, here applied to the initiation of the optimizing data-to-learning-to-action process itself.

Results that matter don't necessarily need to be a full-blown solution to a limiting constraint and associated value-of-learning quantification. Simply demonstrating in a first phase the identification of a limiting constraint that wasn't necessarily previously recognized as such is a good initial deliverable, along with a quantified estimate of what the limitation is costing the organization. Then, in a next phase, you can go further and identify some potential solutions and the associated expected total value of the solutions.

With the optimizing data-to-learning-to-action method, you are ahead of the executive buy-in game because credible quantifications of value are something most executives not only greatly appreciate but also cannot generally ignore—they have a *fiduciary responsibility* to take that seriously. The author has simply never experienced an instance in which senior executives essentially ignored *bona fide* opportunities for value creation that were backed up by *credible* quantifications.

And, again, executive commitment based on talk and theory is only required for taking a *first* step, not *all* the steps. Take the first step, show results, and it will be easy to get buy-in for the next steps. Whether it is formally considered as such or not, it's again an exercise of experimental design at work. Low-cost proof-of-concept learning efforts are committed to first, and as those results are evaluated, if warranted, bigger commitments for enhanced learning are made. Optimizing data-to-learning-to-action projects and initiatives must obey the same methodological prescriptions that the method makes for every other learning and decision opportunity!

Organizing the Initiative

Once you have basic buy-in for at least a first step, it's time to organize your project. There are no magic formulas for that—just rules of thumb that can be adapted to the realities of *your* organization and the particular scope of your initiative.

As with any ambitious initiative, leadership is critical—leadership at the executive-sponsorship level and leadership at the optimizing data-to-learning-to-action working-team level. An executive who "gets it" and is willing to champion the initiative with her peers is key. It is useful to have that individual serve as the chairperson of a *steering team* that the working team reports to. The steering team should include executives who have managerial roles with respect to different areas of the data-to-learning-to-action chain that is being addressed by the initiative such that there is good coverage along all the chain's elements. And even if their organization is not directly involved within the specific chain that is targeted, including an executive from the Finance organization on the steering team can be valuable given the focus the initiative will have on quantifiable value.

The *working team* should be scaled appropriately according to the estimated work that is required for optimizing the target data-to-learning-to-action processes. As mentioned previously, even for the smallest initiative you need a team of at least two people who are familiar with the optimizing data-to-learning-to-action method, with at least one having requisite modeling skills.

For the modeling skills, if you have data scientists or those with similar quantitative skills in your organization, they can be enlisted to help encode required probability distributions and then quantify value by applying modeling techniques such as those that were discussed in Chapter 6. These skills, of course, can also be obtained from outside your organization if necessary.

Also, as mentioned in Chapter 6, there are many techniques and skills for helping people think about uncertainties that can be very valuable to have on the team. That starts with having interviewers who are good listeners and are not overly judgmental, because it is critical to get past the superficial and the "tell-them-what-they-want-to-hear" when conducting interviews. People with an understanding of systems thinking and perspectives on complex adaptive systems are very valuable to have on a data-to-learning-to-action initiative because they can help calibrate against overconfidence in assessing uncertainties.[1] People with such skills can be ideal for conducting the structured interviews, augmented with their own techniques, to get to the encoded probabilities our method needs. What's important, however, is that we get to that *last step*, effectively encoding uncertainties into probabilistic assessments, which is what is so often missing from the various valuable precursor approaches.

As also mentioned previously, it can help to have one or more resources on the team who are *not* familiar with the target data-to-learning-to-action process and the participants in the process. These "beginner minds" are often able to ask the naïve questions that surface insights and opportunities that might otherwise be missed.[2] There also tends to be a psychology at work when colleagues interview colleagues that can suppress questions that need to be asked and topics that need to be probed, and therefore a resource on the team who is outside of the interviewees' immediate work area is helpful.

Further, while familiarity may not necessarily breed contempt, it *can* breed taking colleagues for granted. There are often tremendous insights within an organization that haven't been tapped for just that reason. The author was once conducting an engagement across an organization that included R&D functions, and the author casually referred to a couple of the colleagues in R&D as *research scientists*. These were indeed scientists who had worked for

---

[1]For example, the Cynefin framework can be a helpful start in considering levels of complexity and associated effects on the assessment of uncertainties. See: Kurtz, Cynthia F., and David J. Snowden (2003). "The new dynamics of strategy: Sense-making in a complex and complicated world" (PDF), *IBM Systems Journal*, 42(3): 462–83.

[2]For an example of this sentiment expressed in the context of innovation, see: Katila, Riitta, "Too Many Experts Can Hurt Your Innovation Projects", *Harvard Business Review*, December 7, 2017. https://hbr.org/2017/12/too-many-experts-can-hurt-your-innovation-projects

the company for several decades. Their colleagues actually thought it mildly amusing when I referred to them as the organization's research scientists—they had worked beside them for so long they were now just old "Joe" and "Jim"—their distinction as actual *scientists* had slowly faded over time. And hence, these scientists' perspectives and insights outside a very narrow band of their usual work areas were rarely appreciated and solicited. I sometimes had to remind their colleagues of Joe's and Jim's credentialed expertise, who readily acknowledged it once they took a moment to think past the routine of the past many years.

This is a common *opportunity* when optimizing data-to-learning-to-action processes—*reminding* everyone of the talent within the organization that often becomes taken for granted with the passage of time. This is so key because respect for perspectives, particularly assessment of probabilities, is a critical element of the optimizing data-to-learning-to-action method, and we need colleagues to trust that those subjective probabilistic assessments represent the best possible assessments.

And this certainly also goes for management-level colleagues as well as others. As we all know, management, although perhaps not in words, can take their employees for granted over time, and sometimes need to be reminded of the wonderful capabilities of, and insights available from, the members of their organization. Again, putting things into perspective: if we had the current capabilities of colleagues in artificial neural network form everyone would be extraordinarily impressed with the technology! Perhaps in today's world, that is the way to think of colleagues in order to overcome the tendency to take them for granted.

So, it is critical to have team members who are ready to genuinely learn from the participants who are interviewed, and whom interviewees will feel entirely comfortable providing their honest opinions and assessments. And if you are working on multiple data-to-learning-to-action processes with different sub-teams assigned to each one, it is, of course, important to have excellent communication and working rapport across the teams to ensure, for example, that synergy value is not missed.

Dedicated Resources

When individual projects begin to gain significant mind share within an organization and transform into a sustainable data-to-learning-to-action initiative, it is time to consider establishing a continuing, dedicated part of the organization that is focused on providing data-to-learning-to-action support to the rest of the organization.

Such a dedicated group's responsibilities can include the following:

- Coordinating the organization's overall data-to-learning-to-action initiative
- Communicating the experiences and successes of individual projects, as well as the data-to-learning-to-action initiative at large, across the organization
- Delivering optimizing data-to-learning-to-action *skills* to the rest of the organization
- Providing experienced resources to help *manage* data-to-learning-to-action projects
- *Integrating* individual learning-value diagrams and other data-to-learning-to-action modeling from across the organization
- Assisting in providing oversight of solution implementations and assessments of results

As with other aspects of the data-to-learning-to-action agenda, it makes sense to establish dedicated resources in a step-wise manner, with strong value justifications for each step, coupled with an ongoing assessment of results.

Phasing an Individual Project

As is the case for any significant project, partitioning an optimizing data-to-learning-to-action project into manageable phases is an important success factor. Each phase should be able to be properly resourced and have defined deliverables that can be tracked to gauge success. The following summarizes a typical phasing approach for a specific project.

Phase I:

- Identify the data-to-learning-to-action process and the associated decision to be addressed, supported by value-driver analysis that confirms the target data-to-learning-to-action process's importance to the organization.
- Determine a rough estimate of the upside opportunity of reducing uncertainty related to the decision.
- Estimate the resources required to perform an optimizing data-to-learning-to-action analysis.
- Review results with colleagues and executives and secure commitment for the next phase.

Phase II:

- Work backward from the decision along the data-to-learning-to-action chain and identify the limiting constraint(s) on value. Calculate the value of completely resolving the constraint, which provides an upper bound on the value of any possible solution.

- Evaluate constraint solution options and determine the expected total value of each solution.

- Evaluate synergy value across data-to-learning-to-action processes if appropriate.

- Determine the solution with the greatest positive expected total value.

- Review results with colleagues and executives and secure commitment for implementing the greatest expected total value solution.

Phase III:

- Implement the solution.

- Monitor the results after solution implementation (see the "Management Operating System" discussion later in this chapter).

- Determine if previous modeling should be adjusted based upon an assessment of the results of the implementation.

- If warranted, return to Phase II to determine the next limiting constraint on learning and continue forward with the process for that constraint.

These sample project phases, of course, represent just one possible phasing approach and should be adjusted as required to fit with the realities of a given implementation environment so as to maximize the probability of project success.

And as discussed earlier, for an *initial* optimizing data-to-learning-to-action project for an organization, these phases can be tuned and scaled appropriately so as to constitute a sequence of *proofs of concept* for the application of the method, with, for example, the first phase serving to prove the ability to effectively and credibly quantify the opportunity, the second phase serving to prove the ability to credibly quantify the expected value of a solution, and the third phase serving to assess the results of the implementation of the solution and thereby proving the value capture.

Change Management: What's in It for Me?

As with any other kind of initiative that potentially causes a change in the way people work, there can naturally be some resistance to a data-to-learning-to-action project. Explicitly recognizing that upfront is therefore critical. For change that is expected to be significant, it may be useful to include change-management skills and experience directly on the optimizing data-to-learning-to-action working team, or at least have ready access to these skills when needed.

There is, of course, both a push and a pull aspect to managing change. As discussed in the "Case for Action" chapter (Chapter 1), for people to move to a new state, there must be some level of dissatisfaction with the current state. For executives who have sincerely internalized a fiduciary responsibility for their organization, that dissatisfaction can come in the form of competitive performance metrics that need improving, such as return on assets. For knowledge workers, it may come more in the form making *explicit* for them the factors that are constraining their ability to be most effective, including their ability to make the best possible decisions in the face of uncertainty.

For the *pull* part of the change-management equation, it is useful to consider and address "WIFM"—What's in it for me?—at the very beginning of the project, as it is so fundamental for getting colleagues on board for any type of cooperative initiative, including optimizing data-to-learning-to-action initiatives. The following are some general WIFMs for stakeholders who are likely to be important contributors to the success to your optimizing data-to-learning-to-action initiative.

Senior executives. As the colleagues most directly accountable for the performance of their organization, the WIFM is straightforward. Since optimizing data-to-learning-to-action is all about delivering untapped value, and that value will ultimately translate directly into financial performance metrics, that kind of change is easy for any senior executive to swallow! But as we have discussed, executives need to see results early on, so organizing initiatives to meet that challenge is important.

Finance executives. Finance executives are all about quantifications—hard numbers, in monetary terms. Optimizing data-to-learning-to-action delivers that for investment opportunities that have traditionally relied on less-tangible business cases. It is a method that delivers to finance executives better metrics upon which to base investment decisions, as well as provides a means to communicate value in consistent, quantifiable terms across the entire organization and opportunity spectrum. Therefore, Finance is not only a pretty easy change-management constituency, but can be an outstanding data-to-learning-to-action advocate.

IT organization. The WIFM here is that optimizing data-to-learning-to-action enables a quantification of benefits that can solidly justify investments in technology. It takes the onus for justification of investments in IT from the IT organization and places it where it should be—with the business organizations, which now have a method to determine the *expected total value* of such investments.

Knowledge workers. Knowledge workers consume data and information and transform it into knowledge and actionable learning. The WIFM for knowledge workers includes making their life easier by identifying and debottlenecking constraints along the data-to-learning-to-action process that inhibit their effectiveness. Additionally, an optimizing data-to-learning-to-action program emphasizes encoding knowledge into assessments of uncertainties, often by tapping into insights that knowledge workers have never directly been asked about, a process for which knowledge workers are invariably very grateful.

Data/machine-learning scientists. The WIFM for data scientists and other "quant jocks" is that their scope can not only increase to include their current area of responsibilities, but can also extend to optimizing data-to-learning-to-action modeling. Additionally, they now have *the* method that can credibly quantify *their own value-add* for all that they contribute to the organization's overall learning processes.

Ensure that these WIFMs are understood upfront to facilitate smoothly moving forward, and then be sure to demonstrate with actual examples that your data-to-learning-to-action project has delivered on its promised WIFMs so that you continue to receive maximum support going forward.

Measuring Success

Assessing progress and *value delivery* on a *continuing basis* is critical. A *Management Operating System* for an optimizing data-to-learning-to-action initiative is a means to achieve that continuing assessment. It consists of a process of oversight and an associated set of metrics that provide a continuing overview of progress to stakeholders of the optimizing data-to-learning-to-action initiative. Management Operating System metrics that are valuable to track include the following:

Learning Value. Learning *value*–based metrics are the most important metrics to track and communicate, and learning value can be usefully evaluated both prospectively and retrospectively. Prospectively, the key metric to track is the currently calculated expected total value that is associated with an optimization of a data-to-learning-to-action process. The expected total value can be broken down into its components of expected learning value and the associated investment cost (i.e., the expected direct value) to highlight the *investment leverage*, which is particularly important in budget-constrained environments as discussed in Chapter 7.

Retroactive assessment of learning value can also be useful and instructive. As discussed in Chapter 7, that can be assessed by determining the value that has accrued because different actions were executed versus the actions that would have been executed without the benefit of the learning. Again, it is important that it is understood that this metric does not reflect decision or learning *quality per se*, because decisions cannot be validly evaluated based on outcomes. But, in the aggregate, and with this caveat in mind, retroactive measurement of the value that learning has delivered can be instructive about not only the value that has been delivered, but also as an input to help continuously improve the modeling of uncertainties and value.

Method Value Coverage and Opportunity. This metric pertains to the *coverage* of the various data-to-learning-to-action processes throughout the organization by the optimizing data-to-learning-to-action method, as well as highlights the potential application areas that have not been currently addressed. It is particularly useful to *rank* these areas by expected *value leverage*, which is derived from value-driver analysis, to provide a sense at a glance of the relative potential value *upside* of the areas of the organization that are being currently covered by application of the method versus the potential upside value of future opportunity areas in the organization.

Method Progress. This metric pertains to measuring the progress being made in applying the method for a given data-to-learning-to-action process. Specific milestones that can be tracked include the extent to which learning-value upside (value of perfect decision making) has been determined for the associated decision, the extent to which uncertainties have been encoded into explicit probability distributions, if a limiting learning constraint has been determined, if a preferred solution to the limiting constraint has been determined, and if an expected total value has been explicitly calculated for the preferred solution. Method progress can be a particularly important metric early in an optimizing data-to-learning-to-action initiative before there are significant value results to report.

Learning Quality. At an initial level of evaluation, learning *quality* is an assessment of whether all the appropriate steps of the optimizing data-to-learning-to-action method have been applied. Therefore, it is similar to the Method Progress metrics, except that it is a verification that the steps of the method have been completed for a specific data-to-learning-to-action process rather than a checking of intermediate progress in performing the steps. But, even further, if appropriate, the quality of individual steps of the method can be assessed, such as whether the best possible ability to predict an uncertain variable given a specific state of information has been achieved, or whether all the possible valuable learning variables for a specific decision have been identified. Of course, it generally does not make sense to essentially re-do the original analysis. Rather, assessing whether best practices for these assessed steps were applied is a reasonable check on the learning quality.

These metrics can then be organized into convenient summaries such as dashboards that serve to keep executives' and other stakeholders' attention fully focused on the organization's *learning*, thereby keeping the required resources flowing to high-leverage data-to-learning-to-action optimization opportunities. And, more broadly, these data-to-learning-to-action dashboards can serve to keep the support and engagement levels of all members of the organization high.

Skeptics to Champions

The method in this book is a pretty straightforward and logical approach that most people will readily accept once they have gained a good understanding of it. But there will inevitably be skeptics. The good news is these skeptics can become some of the best champions for optimizing data-to-learning-to-action. Their skepticism will usually be based on a thoughtful perspective, and the key is to feed that thoughtful energy and then work to direct it toward optimizing data-to-learning-to-action success.

The following are some typical categories of skepticism and responses to that skepticism.

We already do that . . .

A natural reaction to any new method that is introduced into an organization is to either reject it outright or to dismiss it by claiming that the method is already being applied—or, at least, that similar results are already being attained. We'll cover the "reject it" reaction in the following sections and focus on "we already do that" here.

Usually, the best way to address the "we already do it" reaction is to apply an "embrace and extend" approach. Simply accept the *perception* that it is already being done, seek to leverage the fine work that has already been performed, and work to see if there is an opportunity to *extend* what has already been done with optimizing data-to-learning-to-action techniques.

The reality is that the optimizing data-to-learning-to-action is sufficiently novel in its *overall* approach that it almost surely is *not* the case that it is already fully being done, but the goal is not to get into a dispute about that. The goal is to achieve the best business performance possible, and that requires *getting started* on an optimizing data-to-learning-to-action program. It can be foolish to be overly doctrinaire—the goal is to improve the organization's performance, not to argue over nuances or terminology that may not much matter. Theoretical arguments rarely win mind-share in organizations, but results generally do.

This embrace-and-extend approach is particularly applicable to the optimizing data-to-learning-to-action method since it is *broader in nature* than other approaches as it addresses decisions and learning most generally. Compare it to, for example, Six Sigma, which is traditionally focused on quality improvement. This quality-oriented focus is likely to be a useful component of a data-to-learning-to-action analysis, but optimizing data-to-learning-to-action may well put Six Sigma–based solutions into a *broader* decision-making and learning context that can then lead to *additional* value generation. Or, as another example of embrace and extend that we have discussed in earlier chapters, there are often existing optimization models within the organization that can be beneficially extended to enable calculations of expected learning value.

Leveraging within an optimizing data-to-learning-to-action initiative work that has already been done and that contributes to the success of the initiative, and then highlighting this contribution to the organization at large, provides excellent prospects for converting these skeptics into becoming true champions for the project.

It's too much work …

The core of the optimizing data-to-learning-to-action method is determining the limiting constraints on learning value and rigorously quantifying the expected value of enhanced learning. That can indeed seem like more work than is involved in an organization's current ad hoc approaches.

Of course, we can appeal to the reality that it is much more likely that the right level of investments will be made on the right kind of learning than with an organization's current approaches. And that may be a powerful argument, particularly for those who believe that there has been an under-investment in the type of learning that most matters to them personally, or to people with a direct responsibility for business performance.

We can also appeal to the fact that the optimizing data-to-learning-to-action method promotes greater overall *efficiency*, as we discussed in the last chapter with the chess analogy, because it serves to more quickly prune away potential action branches that won't add value, and therefore focuses scarce resources on what *really matters*, enabling a *deeper dive* on finding solutions to those areas that truly matter. Without the optimizing data-to-learning-to-action method it can be difficult to sustain this kind of focus because it can be non-obvious as to what is truly most important, and without credible quantifications of value, there is no viable arbiter between disagreements on the matter. Our monetary-based value metric provides the support needed to "just say no" to efforts that, no matter how admirable they *seemingly* are, will not truly add value to the organization and that divert resources and attention from activities that will. When *everything* is important, *nothing* is important—the optimizing data-to-learning-to-action method beneficially serves to *de-clutter* an organization's decisions.

But this is also another case in which the better approach to winning over people to the data-to-learning-to-action method may be through not being overly pedantic. Remember our mantra of "Better to be approximately right than precisely wrong?" Precisely wrong is to bite off too much too soon and then fail to achieve meaningful results, thereby proving the skeptics right. Scale initial efforts appropriately—it is far better to do a first-cut analysis than to get mired down and do nothing at all. Choose your battles well.

The reality, of course, is that everything seems hard until you actually start doing it. Then, it just becomes natural. Common patterns begin to emerge, and people become increasingly efficient at performing the method. Get people *involved* early and often with your optimizing data-to-learning-to-action projects, and their skepticism will likely begin to melt away.

It doesn't apply to the creative parts of the organization . . .

Admittedly, on the surface at least, the data-to-learning-to-to-action method can seem fairly "left-brained."[3] It's a pretty structured approach, and it's also fairly quantitative. Even though it's all about *learning*, which is that traditionally most "scruffy" of concepts. So, while the approach might immediately resonate with, say, engineering types, those in the more people-oriented parts of the organization and most generally, colleagues who tend toward what would be considered more "right-brained" perspectives, are more apt to be skeptical. The optimizing data-to-learning-to-action method may strike them as being too structured, too left-brained, and seemingly constraining on creativity and innovation.

As with the other types of skepticism, it pays to acknowledge the perspectives of the skeptics and agree that they are indeed correct that the method is more structured than the status quo processes. However, it should be emphasized that the additional structure is what enables the *creative energies* of the organization to be resourced and unleashed on what's most *important* for the organization. And the data-to-learning-to-action method itself puts a premium on bringing creativity to bear on developing new actions that have not been

---

[3]While the degree of lateralization of the brain assumed by the original left-/right-brained theory has been shown to be overstated, the labels for describing associated personality traits still often resonate and so are used here.

considered before, and on finding innovative and high-leverage solutions to resolve uncertainties. Innovation techniques such as *design thinking* can be very helpful in this regard, for example.[4]

But that's just so much talk—it really needs to be *demonstrated* to them that all the creative power of the organization is encouraged and leveraged by the optimizing data-to-learning-to-action method. And that optimizing data-to-learning-to-action is simply aimed at getting the *most* out of the creative capabilities by directing them toward the most important aspects of the organization and then encoding the resulting insights so that they can be most usable for decision making. This is another case of "seeing is believing," so getting those with this brand of skepticism involved early and often is key. And after seeing the premium that the method, in practice, places on innovative thinking, they will likely become strong advocates for the method.

And for the more philosophically inclined skeptics, a sub-objection that is a variation of the following may be articulated, warranting a deeper dive with them into the underpinning perspectives of the optimizing data-to-learning-to-action method.

Learning *must* be more than just reducing uncertainty. What about the *pure exploration* of new domains?

While people generally concede that most learning, at its core, is a process for reducing uncertainty to be able to better predict, it may be more difficult to accept that this applies even for pure discovery-type learning. But the perspective that the learning process is essentially a process of reducing uncertainty can indeed be considered to encompass exploration and discovery as well. That's because even if a completely new domain of understanding has not even yet been identified, it can be argued that by the act of simply *identifying* the domain, reduction in uncertainty occurs, and hence learning, by our definition, occurs.

The differing views on uncertainty seemingly stem from two fundamentally different perspectives on learning and discovery:

- Knowledge and inventions don't exist until we create them.

- All knowledge and inventions exist in a meta-reality; we explore this meta-reality "search space" to find them.

[4]Plattner, Hasso, Meinel, Christoph, and Leifer, Larry J., eds. *Design Thinking: Understand-Improve-Apply (Understanding Innovation)* (Springer-Verlag, 2011).

It seems that either perspective is valid and is not provable to be correct, but the optimizing data-to-learning-action approach implicitly aligns with the second perspective. That's because the perspective that learning is the exploration of a search space more naturally leads to a quantified approach compared to the first perspective. With the concept of the meta-reality search space, each exploration or discovery (i.e., learning) clearly results in a *reduction of uncertainty* with respect to the *overall search space*.

An analogy to this perspective might be Shannon's information theory, which many people resisted as embodying too constraining a concept of information, but that then enabled a rigor, a means for quantifications, and resulting practical applications that would not otherwise be possible.

Pragmatically, these philosophical differences on the nature of learning and uncertainty don't much matter for real-world applications. At the extreme of purely imaginative exploration, it is acknowledged that the method outlined in this book by itself isn't necessarily going to be as *sufficient* (although I argue is still *necessary*) as is the case for most domains of learning in an organization. There are certainly other complementary techniques that should be brought to bear for such learning extremes; for example, design-thinking techniques that were mentioned previously.

To summarize the skeptics-to-champions approach, it should be recognized that skeptics don't blindly follow. They are often the deep thinkers. Take the time to listen and work with them rather than ignoring them, and incorporate their perspectives. It will pay off.

Summary

In this chapter, we discussed organizing approaches for ensuring successful data-to-learning-to-action initiatives. We covered scoping the project or initiative, gaining executive commitment, organizing resources, and phasing the initiative. We reviewed the basics of a management operating system to assist in managing the initiative both before and after solution implementations, including some of the key metrics that should be tracked to gauge ongoing success. We also discussed overall change-management issues, as well as how to redirect skeptics of optimizing data-to-learning-to-action to becoming champions of the method.

Conclusion

We began our optimizing data-to-learning-to-action journey by considering the defining characteristics of this era of business, the unprecedented advances across such a broad front of technologies, and the resulting complexity for business and IT strategies. And, at the same time, we looked at the sobering long-term trends of business-performance metrics such as returns on assets and corporate topple rates.

We concluded that, inevitably, the roads to business-performance improvement, as well as the roads to managing technology-driven complexity and the resulting firehose of data that comes along with it, all lead us back to *decisions*. That ultimately, business performance is most fundamentally a function of the *effectiveness of our decisions* and their implementations, and that the effectiveness of our decisions rests squarely on the effectiveness of our *learning*.

But what exactly *is* learning? We tamed that most slippery of concepts by concluding that, at its core, learning is the ability to better predict by decreasing uncertainty. That holds for learning performed by minds. That holds for learning performed by machines. So, learning is the process of reducing uncertainty. But it can also be considered a *flow* along a process. Specifically, the data-to-learning-to-action process, which is the *universal* process for learning. Universal because it encompasses nearly everything of importance that we do. Universal because birds do it, we do it, even educated machines now do it.

And with the perspective that learning is a process for reducing uncertainty, we found that we could quantify the value of *actionable* learning—learning that has the potential to affect a decision. We could put a monetary value on this actionable learning and thereby effectively evaluate investments in

S. Flinn, *Optimizing Data-to-Learning-to-Action*,
https://doi.org/10.1007/978-1-4842-3531-7_11

learning on the same credible basis as every other investment opportunity. The value of learning need no longer be solely the province of the feel-good pronouncements and hand-waving value propositions that simply may not get it right, and even if they do *happen* to get it right, will likely fail to provide a basis for *sustained* focus.

With a quantified upper bound on learning value as a guide, we could work backward along a data-to-learning-to-action process and identify constraints in the flows of learning. And, specifically, we could identify a *limiting* constraint on learning value and quantify the expected value of partially or fully resolving the limiting constraint by applying people, process, and/or technology-based solutions.

When that limiting constraint was resolved, the associated decision would be better than before, and additional, tangible value would therefore be delivered. There would also then *necessarily* be *another* limiting constraint on learning value to be addressed. The process of resolving these limiting constraints continues until the expected value of resolving the next limiting constraint no longer exceeds its expected costs.

This net value we call the *expected total value*, and total value applies more broadly to *any* investment opportunity and therefore provides the proper basis for optimizing an organization's overall portfolio of investment opportunities. And that leads to improved business performance on a sustained basis.

That was the basic journey we took. Let's now review the core concepts of the book in a bit more detail. Then, we will close by considering why optimizing data-to-learning-to-action represents the *next era* of business-performance improvement, and then lay down the challenge of taking the next steps to kick off that era in *your* organization.

Core Concepts

The following is an overview of the key concepts of the book, roughly in the order of their introduction and their logical flow.

Business performance must be improved. And with new methods. Notwithstanding the veneer of, for example, a rising stock market, more fundamental measures of business health such as returns on assets for US companies have been trending downward for decades. Shareholder value is being destroyed by the lower quartile of businesses. Topple rates of businesses have generally been increasing. Traditional performance-improvement approaches have clearly not been sufficient, and they can be expected to be even less sufficient for the future business environment.

Business performance is driven by the effectiveness of decision making. Decision effectiveness and financial results correlate at a 95 percent confidence level. The inevitable conclusion is that the business underperformance that has led to declines in metrics such as return on assets is therefore necessarily driven by systemic issues related to business decision making. And, of course, that conclusion comports with good old-fashioned common sense.

Decision making is particularly challenging in the face of increased complexity and uncertainty. Increasing global competition and the unprecedented acceleration of technology advances make for a highly challenging decision-making environment. Ironically, while technology-based progress should be good news and be part of the performance-improvement equation, without benefit of best practice decision-making processes, the complexity factor associated with the accelerating technology advances, along with the waves of additional data that are being generated, risks resulting in business performance that is actually worse than it otherwise would be.

Decisions that are directly tied to value drivers have the greatest leverage on business performance. Value drivers are determined by breaking down an organization's financial statements into meaningful components and performing sensitivity analysis to determine the factors that have the greatest financial leverage. Improving the decisions that are directly associated with these highest-leverage factors is therefore the key to improving business performance.

It's not about data—it's about decisions. And learning. Data has no *intrinsic* value—data only has value to the extent that it can affect decisions. Fixating on data, no matter how big the data, without considering its direct connection to decisions is the stuff of Dilbert cartoons and is a prescription for value destruction. Learning is the *bridge* between data and decisions, and learning that can potentially serve to change a decision has tangible value.

Learning is the process for making better predictions through the reduction of uncertainty. Learning, explicitly or implicitly, results in the shrinking of probability distributions, thereby increasing predictive accuracy. Predictive accuracy directly affects the efficacy of decisions, which, again, directly affects business performance.

Learning that can potentially affect decisions is actionable learning, and actionable learning has a tangible, quantifiable expected value. Expected learning value is analogous to the concept of value of information in the field of decision analysis. But information, like data, does not have *intrinsic* value—it is the learning that beneficially makes use of the information that has positive, quantifiable value if the learning has the potential to change a decision.

The expected value of learning is the difference between the expected value of a decision with benefit of the learning and the value of the decision without benefit of the learning. This is the fundamental prospective value of learning formula and it always holds true. The way we calculate the expected value of the decision with or without benefit of the additional learning may be based upon various techniques such as decision trees or simulations—ultimately, all boiling down to the application of Bayesian methods. But taking the difference between the two prospective decision situations—with and without benefit of additional learning—is then always performed to derive the expected value of the additional learning.

Data-to-learning-to-action is a process. In fact, it is the *universal* process, because it is the fundamental process for learning and decision making, whether performed by minds or machines. Data-to-learning-to-action processes are therefore pervasive throughout any organization.

Data-to-learning-to-action processes can be segmented into a set of standard elements. A robust segmentation of elements that serves well for most optimizing data-to-learning-to-action applications in organizations includes the following sequence of elements: Data Acquisition, Data Filtering, Information Management, Search and Discovery, Predictive Analytics, Process and Collaborate, and Decide and Act. Each of these elements typically includes both people- and technology-based capabilities and activities.

Learning-value diagrams can be beneficially applied to model data-to-learning-to-action processes. Learning-value diagrams are derivatives of decision diagrams. They model decisions, uncertainties, learning variables, and associated relationships, and can be extended and refined as an optimizing data-to-learning-to-action project or initiative progresses. The learning variables represent the actionable learning opportunities for reducing uncertainty and improving decisions.

All processes have a flow and a limiting constraint on the flow throughput. That's the fundamental law of the Theory of Constraints. And since data-to-learning-to-action is a process, it necessarily has a limiting constraint on its throughput. The limiting constraint of a data-to-learning-to-action process is a constraint on the *flow of learning*.

The limiting learning constraint of a data-to-learning-to-action process is determined by working backward from decision-to-data. Systematically working backward from the decision and the uncertainties that affect the decision along a flow of learning of the data-to-learning-to-action process enables identification of the limiting constraint of the learning flow. Techniques such as structured interviews are applied to facilitate the identification of the limiting constraint.

Resolving the limiting learning constraint is mandatory to enable the data-to-learning-to-action process to generate more value. This fact also follows from a fundamental law of the Theory of Constraints: learning *throughput* can *only* be increased by resolving the *limiting* learning constraint. And the corollary is that investing in learning capacity elsewhere in the data-to-learning-to-action process without addressing the limiting constraint of the flow is futile and actually serves to destroy value.

Limiting constraints are addressed by people-, process-, and/or technology-based solutions. Solutions to address the limiting constraints are aimed at reducing one or more uncertainties that affect the associated decision—the solutions therefore directly correspond to learning variables in learning-value diagrams. It most typically requires *combinations* of people-, process-, and/or technology-based enhancements to most effectively address the limiting constraints.

Determining solutions to limiting learning constraints is an exercise in creativity. While the data-to-learning-to-action method provides for additional structure versus more ad hoc approaches, this structure serves to identify the right place to focus an organization's creative powers, such as finding the best possible solutions to limiting learning constraints, thereby *amplifying* the value of the creativity.

The value of partially or fully resolving a limiting constraint of a data-to-learning-to-action process can be calculated. The people-, process-, and/or technology-based solution that can be potentially applied to the limiting constraint serves to reduce uncertainty, which enhances predictive accuracy and thereby improves decision making. The calculated *expected learning value* of such a solution is a function of the degree to which the uncertainty is reduced, the uncertainty's leverage on a decision, and the decision's leverage on financial performance. There may also be synergy value across multiple data-to-learning-to-action processes that contributes to the expected learning value of a solution.

Alternative solutions to limiting learning constraints are evaluated based upon their expected total value. Expected total value includes the expected learning value of a solution calculated on a discounted cash-flow basis *net* of the expected cost of the solution.

Expected total value equals expected learning value plus expected direct value. Expected direct value is defined as the standard expected discounted cash flow of a project or activity in which uncertainties are taken as a *given*. It also includes the investment costs that are associated with learning, which aims to *reduce* the currently *given* uncertainties. Expected learning value essentially converts to expected direct value after the learning occurs, with the uncertainties updated by the learning process becoming the new givens. This conversion process represents the defining dynamic of the *business-renewal* lifecycle.

Expected total value should be used to prioritize an organization's portfolio of investment opportunities. The expected total value of some of the projects and activities of an organization will primarily comprise expected learning value, some primarily expected direct value, and some a significant mix of both. Regardless of the mix, if the expected total value of a project or activity is *positive*, it should be implemented. Alternatively, in budget-constrained environments, the portfolio of investment opportunities should be optimized to *maximize expected total value* of the portfolio subject to the budget constraints.

Implement and learn. Organize to successfully optimize data-to-learning-to-action processes, identify the highest-leverage limiting constraints on learning, implement the solutions with the greatest expected total value that resolve the limiting constraints on learning, monitor and measure the business-performance results, learn from the implementations, and re-calibrate models and uncertainties based on the learning for the next iteration of applying the optimizing data-to-learning-to-action process.

The Data-to-Learning-to-Action Era

In the book, *The Learning Layer*, I proposed that we were entering a fundamentally new era of information technology, an era in which systems automatically adapt by learning from us and then continuously deliver the results of that learning back to us to make us smarter and more productive. And that era is clearly upon us, although as marvelous as the advances of the past few years have been, it is clearly still only at its very beginning.

And as marvelous as the advances have been and will be, as with any technology advances, they certainly do not guarantee improved business performance. In fact, as we have seen, it can be just the opposite at both a macroeconomic level as well as for any particular organization. For improved business performance, we need something more. And that something more is the optimizing of data-to-learning-to-action processes.

We have had an analogous business situation before. As we entered the 1990s, Western businesses were feeling a persistent competitive squeeze. Globalization had begun in earnest, and the bar on efficiency and product quality had suddenly been raised. Financial results waned, and plants closed.

There was a lag, but businesses eventually reacted. Both product quality and efficiency came under an intense focus for improvement. It set off a continuing era of business-performance improvement that can be labeled the "Reengineering era"—the last great wave of business-performance improvement. This was a highly successful era for businesses that truly embraced the required analysis and change; although as we know from the macro-level data, more generally, it wasn't enough.

Now we are in need of a *new* business performance–improvement era, and I believe that era is necessarily the optimizing data-to-learning-to-action era, because only it embodies an approach that both addresses the core issues of business performance as well as fully embraces the promises of the coming era of increasingly intelligent technology.

A comparison of the last great business performance–improvement era and the emerging one that I am now advocating is illustrated by Figure 11-1, which summarizes the basic similarities and distinctions between the reengineering era and the upcoming optimizing data-to-learning-to-action era along some key dimensions.

As we walk through Figure 11-1, notice the *parallels*, as well as the *contrasts*, with the reengineering paradigm of the 1990s–2000s. For example, in the 1980s, Japan was the new international-based competitive threat that ultimately drove not only the reengineering wave, but also the increase in the popularity of methods such as quality improvement and constraint theory, all of which ultimately contributed toward significantly stronger businesses and practices around the world.

	Reengineering Era	**Data-to-Learning-to-Action Era**
International Driver	Japan	China
Technology Driver	ERPs	Cloud/Big Data/IOT/ML/AI/BI/Collaboration etc.
Financial Driver	Efficiency	ROA/Total Value
Process Focus	Transactions/Operations	Learning/Decisions
Era	1990-2010	Next 10+ Years

Figure 11-1. Reengineering versus Optimizing Data-to-Learning-to-Action Eras

Now, the ascendant international competitive threat is clearly China, and unlike Japan, China has the size advantage and momentum to be more of a sustainable driver of competitive pressure. And not just limited to primarily raising the competitive bar in the manufacturing sector as was the case of Japan, but in just about every sector of the economy, including the most advanced areas of both theory and application in such scientific fields as AI, biotech, and materials science.

In the now seemingly quaint technology environment of the 1980s and 1990s, the driver of much of the reengineering wave was enterprise resource planning (ERP) systems. These systems, at least initially, were primarily focused on manufacturing and transactional processes such as procurement. That was the disruptive enterprise technology of that era.

Now, of course, we have the cloud, internet of things, AI/machine learning, advanced business intelligence capabilities, new ways of collaborating, and so forth. And right behind these technologies are blockchain, virtual/mixed realities, 3D printing, and so on. Just as China is clearly a more competitively disruptive force of globalization than Japan could ever be, the disruptive potential of today's technologies compared to that of at least the early part of the reengineering era may well be on the order of a magnitude greater.

The financial driver of the reengineering era was primarily efficiency, which is certainly still important. However, this prevailing focus of reengineering is also why it was often greeted with less than enthusiasm by many employees, as it was sometimes viewed as merely a euphemism for workforce reduction. In contrast, the financial drivers for optimizing data-to-learning-to-action are more holistic. From an overall business-performance perspective, optimizing data-to-learning-to-action is aimed at improving return on assets, among other important financial metrics. And from an individual investment-opportunity perspective, as well as for portfolios of investment opportunities, it is focused on total value, which explicitly includes the value of learning. So, it touches *every* economic decision in an organization.

Whereas reengineering was oriented more toward improving transactional and operational areas of the organization, optimizing data-to-learning-to-action has the broader remit of being directed at improving learning and decisions generally, impacting every area of an organization, including the *core* areas of competitive advantage for just about every business model— namely, the processes through which the organization continually *grows and renews* itself.

And because optimizing data-to-learning-to-action is more holistic in its business-performance goals, as well as with respect to its scope of coverage of the functions and processes of an organization, it is not prone to the counter-productive sub-optimizations and other economic distortions that can be such a common consequence of more-limited focuses on performance improvement.

How long is the data-to-learning-to-action era destined to prevail? Certainly, advances in technology will in no way obviate its importance. On the contrary, the advances will only make optimizing data-to-learning-to-action more critical, particularly as machine-based learning becomes ever more prevalent and powerful.

So, it is hard to envisage a time in which optimizing an organization's data-to-learning-to-action processes will *not* be the most relevant lever for sustainable performance improvement. After all, it *is* the *universal* process that is being optimized, the process for learning. What could be more fundamental?

Next Steps

So, what's next for you and your organization? I would suggest trying out an approach that

- provides a way to successfully navigate *today's* highly disruptive competitive and technology environment;

- focuses on the *core* of sustainable competitive advantage—improved learning and decision making, including learning and decision making for the most *strategic* areas of the business;

- supports sustained organizational *renewal* by credibly quantifying the value of investments in learning;

- *de-clutters* an organization's decision making;

- delivers direct improvement of *key* financial indicators; and

- is fully applicable to *all* parts of an organization.

These are, of course, a combination of attributes that are unique to the optimizing data-to-learning-to-action approach. And they should also be aspects that check the boxes that are required to get the attention of decision makers who can help you make it a reality for your organization. Chapter 10 has some suggestions on how to get started.

So, this book has provided you with data and information. From this data and information, you have learned. And you know by now that the next step of the process is . . . act!

I

Index

A

AlphaZero, 158

Anticipated capabilities, 129–131

Application programming interfaces (APIs), 9, 50

Artificial neural networks, 35, 43, 166

B

Bayesian approach, 103–104

Biological-based neural networks, 35

Business intelligence systems, 52–53

Business-renewal lifecycle
action intention matrix, 113
autopilot, 114–115
description, 109
expected direct value, 110–111
expected learning value, 110–111
experience-curve effects, 113–114
experimental design, 112
graduates, 111
learning-value diagrams, 115–116
manufacturing techniques, 114
operational decisions, 115
processes and activities, 115
projects and activities, 111
R&D and pilots, 111, 112, 114
software-related jobs, 112
type I and type II errors, 112–113

C

Cloud computing models, 8

Common learning constraint, 126

Constraint look-ahead, 127–129

Customer relationship management (CRM), 56–57

D

Data Acquisition element, 126, 128, 132, 154–156

Data clustering, 43

Data Filtering element, 128, 132, 153–154

Data-to-learning-to-action
process, 14–15, 26–28
anticipatory computing, 36–38
APIs, 50
architecture of learning, 35
artificial neural network, 166
assessment of probabilities, 166
beginner minds, 165
biological-based processes, 31
business and IT strategies, 177
business intelligence, 52–53
business-performance improvement, 61, 177, 183
change management, 169–170
commonly-understood-to-need-improvement factor, 162
company's financial performance, 69
computer-based forms, 27

Get the eBook for only $5!

Why limit yourself?

With most of our titles available in both PDF and ePUB format, you can access your content wherever and however you wish—on your PC, phone, tablet, or reader.

Since you've purchased this print book, we are happy to offer you the eBook for just $5.

To learn more, go to http://www.apress.com/companion or contact support@apress.com.

Apress®

Printed in the United States
By Bookmasters